Proposed Carbon Pollution Emission Guidelines for Existing Electric Utility Generating Units

Introduction

The purpose of this memorandum (Legal Memorandum) is to supplement the preamble by providing background for the legal issues discussed in the preamble for this proposed rule[1] and further discussion of some, but not all, of those issues. This memorandum is intended to be read in conjunction with, and assumes familiarity with, the preamble.

I. Background

A. Clean Air Act section 111

Clean Air Act (CAA) section 111, which Congress enacted as part of the 1970 CAA Amendments, establishes mechanisms for controlling emissions of air pollutants from stationary sources. This provision requires EPA to promulgate a list of categories of stationary sources that the Administrator, in his or her judgment, finds "causes, or contributes significantly to, air pollution which may reasonably be anticipated to endanger public health or welfare."[2] EPA has listed more than 60 stationary source

---

[1] The proposed rule is the "Carbon Pollution Emission Guidelines for Existing Stationary Sources: Electric Utility Generating Units."

[2] CAA section 111(b)(1)(A).

categories under this provision.[3] Once EPA lists a source category, EPA must, under CAA section 111(b)(1)(B), establish "standards of performance" for emissions of air pollutants from new sources in the source category.[4] These standards are known as new source performance standards (NSPS), and they are national requirements that apply directly to the sources subject to them.

When the EPA establishes NSPS for new sources in a particular source category, the EPA is also required, under CAA section 111(d)(1), to prescribe regulations for states to submit plans regulating existing sources in that source category for any air pollutant that, in general, is not regulated under the CAA section 109 requirements for the national ambient air quality standards (NAAQS) or regulated under the CAA section 112 requirements for hazardous air pollutants (HAP). In contrast with CAA section 111(b), which provides for direct federal regulation of new sources, section 111(d)'s mechanism for regulating existing sources provides that states will submit plans that establish "standards of performance" for the affected sources and that contain other measures to implement and enforce those standards.

---

[3] *See* 40 CFR 60 subparts Cb – OOOO.
[4] CAA section 111(b)(1)(B), 111(a)(1).

The term "standard of performance" is defined under CAA section 111(a)(1) as a "standard for emissions of air pollutants" that "reflects the degree of emission limitation achievable" from the "best system of emission reduction," considering costs and other factors, that "the Administrator determines has been adequately demonstrated." CAA section 302(l) also defines "standard of performance" as "a requirement of continuous emission reduction, including any requirement relating to the operation or maintenance of a source to assure continuous emission reduction."

Under the EPA's implementing regulations for CAA section 111(d)(1), the EPA must determine the best system of emission reduction for the sources, and then apply that best system to determine the required level of emissions or emission reduction, which the regulations refer to as the "emissions guideline."[5] Under section 111(d)(1), the states must then adopt state plans that establish standards of performance and measures that implement and enforce those standards. In the case of an air pollutant that EPA has determined may cause or contribute to endangerment of public health, the states' standards of performance must

---

[5] 40 CFR 60.22(b)(5).

not be less stringent than the EPA's emission guideline.[6]

CAA section 111(d)(1) grants states the authority, in applying a standard of performance to particular sources, to take into account the source's remaining useful life or other factors.

The state must submit its plan to the EPA for approval, and, under CAA section 111(d)(2), the EPA must approve the state plan if it is "satisfactory."[7] If a state does not submit a plan, the EPA must establish a federal plan for that state.[8] Once a state receives the EPA's approval for its plan, the provisions in the plan become federally enforceable against the entity responsible for noncompliance, in the same manner as the provisions of an approved state implementation plan (SIP) under CAA section 110.

B. Legislative history

The legislative history of the 1970 Clean Air Act Amendments indicates that at that time, Congress grouped air pollutants from existing stationary sources into three categories: (i) air pollutants that affected the National Ambient Air Quality Standards (NAAQS), which would be regulated under CAA section 110 state implementation plans

---

[6] 40 CFR 60.24(c).
[7] CAA section 111(d)(2)(A).
[8] *Id.*

(SIPs), (ii) hazardous air pollutants (HAPs), which would be regulated under EPA-promulgated national emission standards pursuant to CAA section 112, and (iii) all other air pollutants. The House bill did not address this third group of air pollutants, but the Senate bill did: it termed them "selected air pollution agents" and proposed to require the EPA to promulgate national emission standards pursuant to proposed CAA section 114. The 1970 House-Senate Conference Committee that was formed to resolve differences between the House and Senate versions of the CAA Amendments did not adopt the Senate bill's proposed CAA section 114, but did adopt section 111(d), which covers the same non-NAAQS, non-HAPs air pollutants. Under section 111(d)(1) as included in the 1970 CAA Amendments, the states were required to submit to the EPA state plans that "establish[] emission standards" for their existing sources. Although the legislative history of the 1970 CAA Amendments does not contain statements that directly discuss the specific provisions included in section 111(d), the legislative history of the Senate bill's proposed section 114 is relevant to the meaning of section 111(d), and we refer to parts of that legislative history below.

For new sources, section 111(b)(1)(B) required the EPA to promulgate "standards of performance," and defined that term, under section 111(a)(1), as—

> a standard for emissions of air pollutants which reflects the degree of emission limitation achievable through the application of the best system of emission reduction which (taking into account the cost of achieving such reduction) the Administrator determines has been adequately demonstrated.

The legislative history discusses, among other things, the meaning of the term "standard of performance,"[9] which we refer to below.

In the 1977 CAA Amendments, Congress made several changes to section 111, including section 111(d). Congress substituted "standards of performance" for "emission standards," which, as noted above, the states are required to establish in their state plans. In addition, Congress added to section 111(d)(1) the requirement that the EPA's regulations "permit the State in applying a standard of performance to any particular source under a [section 111(d)] plan … to take into consideration, among other factors, the remaining useful life of the existing source to which such standards applies." Congress added to section 111(d)(2) a similar requirement applicable to federal plans. In addition, Congress revised the definition of

---

[9] *See, e.g.,* Senate Comm. Rep. No. 91-1196 at 16.

"standard of performance" in section 111(a)(1) to distinguish among different types of sources, and to require that for fossil fuel-fired sources, the standard (i) be based on, in lieu of the "best system of emission reduction … adequately demonstrated," the "best technological system of continuous emission reduction … adequately demonstrated;" and (ii) require a percentage reduction in emissions. In addition, in the 1977 CAA Amendments, Congress expanded the parenthetical requirement that the Administrator consider the cost of achieving the reduction to also require the Administrator to consider "any nonair quality health and environment impact and energy requirements." Congress also added the definition of "standard of performance" in section 302(l), which defines the term to require a "continuous emission reduction."

In the 1990 CAA Amendments, Congress made further amendments to section 111, including section 111(d). Among other things, Congress again revised the definition of "standard of performance" under CAA section 111(a)(1), this time repealing the requirements that the standard of performance be based on the best technological system and achieve a percentage reduction in emissions, and replacing those provisions with the terms used in the 1970 CAA Amendments' version of section 111(a)(1) that the standard of performance be based on the "best system of emission reduction … adequately demonstrated." In addition, in section 111(d)(1)(A)(i), Congress revised the description of which air pollutants are subject to section 111(d) but, as discussed below, left the provision ambiguous with respect to its applicability to the air pollutant emitted from the sources at issue in this rulemaking: CO_2 emissions from fossil fuel-fired EGUs . CAA section 111 has not been revised since the 1990 CAA Amendments.

C. Regulatory history and case law

The EPA issued regulations implementing CAA section 111(d) in 1975,[10] and has revised them in the years since.[11] (We refer to the regulations generally as the implementing regulations.) These regulations provide that, in promulgating requirements for sources under CAA section 111(d), the EPA first develops regulations known as "emission guidelines," which establish binding requirements that states must address when they develop their plans.[12]

---

[10] "State Plans for the Control of Certain Pollutants From Existing Facilities," 40 FR 53,340 (Nov. 17, 1975).

[11] The most recent amendment was in 77 Fed. Reg. 9304 (Feb. 16, 2012).

[12] 40 CFR 60.22. In the 1975 rulemaking, the EPA explained that it used the term "emissions guidelines" – instead of emissions limitations – to make clear that guidelines would not be binding requirements applicable to the sources, but instead are "criteria for judging the adequacy of State plans." 40 Fed. Reg. at 53,343.

The implementing regulations also establish timetables for state and EPA action. The default rule is that states must submit state plans within nine months of the EPA's issuance of the guidelines,[13] but the regulations provide the EPA with authority to extend the deadlines for those submissions.[14] The regulations also provide that the EPA must take final action on the state plans within four months of the due date for those plans.[15] In the present rulemaking, the EPA is following the requirements of the implementing regulations, except that the EPA is extending certain timetables, as described in the preamble.[16]

Over the last forty years, under CAA section 111(d), the agency has regulated four pollutants from five source categories (i.e., phosphate fertilizer plants (fluorides), sulfuric acid plants (acid mist), primary aluminum plants (fluorides), Kraft pulp plants (total reduced sulfur), and municipal solid waste landfills (landfill gases)).[17] In

---

[13] 40 CFR 60.23(a)(1).
[14] See id.; 40 CFR 60.27(a).
[15] 40 CFR 60.27(b).
[16] The EPA is not re-opening the existing regulations, although it is revising the deadline for action on state plan submittals. The EPA is proposing additional regulatory requirements, which are contained in proposed subpart UUUU.
[17] See "Phosphate Fertilizer Plants; Final Guideline Document Availability," 42 Fed. Reg. 12,022 (Mar. 1, 1977); "Standards of Performance for New Stationary Sources; Emission Guideline for Sulfuric Acid Mist," 42 Fed. Reg. 55,796 (Oct. 18, 1977); "Kraft Pulp Mills, Notice of

addition, the agency has regulated additional pollutants

under CAA section 111(d) in conjunction with CAA section

129.[18] The agency has not previously regulated CO_2 or any

other greenhouse gas under CAA section 111(d) (although

because landfill gases include methane, the agency's

regulation of landfill gases reduced emissions of that

greenhouse gas).

The D.C. Circuit has never handed down a decision that

interpreted, or reviewed EPA's interpretation of, section

111(d). The D.C. Circuit has, however, reviewed

rulemakings under CAA section 111 on numerous occasions

during the past four decades, handing down decisions dated

from 1973 to 2011.[19] These decisions concerned various

aspects of section 111, primarily the interpretation of the

Availability of Final Guideline Document," 44 Fed. Reg.
29,828 (May 22, 1979); "Primary Aluminum Plants;
Availability of Final Guideline Document," 45 Fed. Reg. 26,294 (Apr. 17, 1980);
"Standards of Performance for New Stationary Sources and
Guidelines for Control of Existing Sources: Municipal Solid
Waste Landfills, Final Rule," 61 Fed. Reg. 9905 (Mar. 12,
1996).
[18] *See, e.g.,* "Standards of Performance for New Stationary
Sources and Emission Guidelines for Existing Sources:
Sewage Sludge Incineration Units, Final Rule," 76 Fed. Reg.
15,372 (Mar. 21, 2011).
[19] *Portland Cement Ass'n v. Ruckelshaus,* 486 F.2d 375 (D.C.
Cir. 1973), *cert. denied,* 417 U.S. 921 (1974); *Essex
Chemical Corp. v. Ruckelshaus,* 486 F.2d 427, (D.C. Cir.
1973), *cert. denied, Appalachian Power Co. v. EPA,* 416 U.S.
969 (1974); *Portland Cement Ass'n v. EPA,* 665 F.3d 177
(D.C. Cir. 2011).

term "standard of performance." Relevant aspects of these cases are discussed below.

D. Summary of section 111 proposals

The EPA is in the process of conducting three rulemakings to regulate CO_2 from fossil fuel-fired electricity generating units (EGUs), including both fossil fuel-fired electric utility steam generating units and natural gas-fired stationary combustion turbines (affected sources or affected EGUs). The first, published in January, 2014, proposes standards of performance under CAA section 111(b) for affected sources undertaking new construction. The second is the present rulemaking, under CAA section 111(d), which proposes emission guidelines for states to follow in adopting state plans that regulate existing affected EGUs. In the third rulemaking, which we expect to propose concurrently with the present one, the EPA is proposing standards of performance under section 111(b) for affected EGUs that undertake modifications or reconstructions.

II. Summary of legal basis

The following summarizes the main features of the EPA's legal rationale for this proposed rulemaking. All of this rationale is discussed in the appropriate sections of the preamble for this rulemaking. This Legal Memorandum elaborates on some, although not, all of these features.

Today's proposed action is consistent with the requirements of CAA section 111(d) and the implementing regulations. As an initial matter, the EPA reasonably interprets the provisions identifying which air pollutants are covered under CAA section 111(d) to authorize the EPA to regulate CO_2 from fossil fuel-fired EGUs. Specifically, an ambiguity in the provisions of section 111(d)(1)(A)(i), arising from Congress's simultaneous enactment of two separate versions of this provision, has led some stakeholders to argue that the fact that the EPA has regulated hazardous air pollutants from EGUs prevents the EPA from regulating CO_2 emissions from EGUs. As explained below, however, the EPA reads the provision to authorize regulation of CO_2 emissions from EGUs and this interpretation is both reasonable and entitled to deference.

In addition, the EPA recognizes that CAA section 111(d) applies to sources that, if they were new sources, would be covered under a CAA section 111(b) rule. The EPA intends to complete two CAA section 111(b) rulemakings regulating CO_2 from new fossil fuel-fired EGUs and from modified and reconstructed fossil fuel-fired EGUs before it finalizes this rulemaking, and either of those section 111(b) rulemakings will provide the requisite predicate for this rulemaking.

A key step in promulgating requirements under CAA section 111(d) is determining the "best system of emission reduction ... adequately demonstrated" (BSER). In promulgating the implementing regulations, the EPA explicitly stated that it is authorized to determine BSER;[20] accordingly, in this rulemaking, the EPA is determining BSER.

The EPA is proposing two alternative approaches for the "best system of emission reduction ... adequately demonstrated" for fossil fuel-fired EGUs, each of which is based on methods that have employed for reducing emissions of air pollutants, including, in some cases, CO_2, from these sources. The first identifies the combination of the four building blocks as the BSER. These include operational improvements and equipment upgrades that the coal-fired steam-generating EGUs in the state may undertake to improve their heat rate (building block 1) and increases in, or retention of, zero- or low-emitting generation, as well as

---

[20] The EPA is not re-opening that interpretation in this rulemaking.

measures to reduce demand for generation, all of which,
taken together, displace, or avoid the need for, generation
from the affected EGUs (building blocks 2, 3, and 4). All
of these measures are components of a "system of emission
reduction" for the affected EGUs because they either
improve the carbon intensity of the affected EGUs in
generating electricity or, because of the integrated nature
of the electricity grid and the fungibility of electricity
and electricity services, they displace or avoid the need
for generation from those sources and thereby reduce the
emissions from those sources. Moreover, those measures may
be undertaken by the affected EGUs themselves and, in the
case of building blocks 2, 3, and 4, they may be required
by the states.

Further, these measures meet the criteria in CAA
section 111(a)(1) and the case law as the "best" system of
emission reduction because, among other things, they
achieve the appropriate level of reductions; they are of
reasonable cost, including when viewed through a nation-
wide lens; they are consistent with trends in the energy
sector; and they encourage technological development and
expansion that is important to achieving further emission
reductions. Moreover, the measures in each of the building
blocks are "adequately demonstrated" because they are each

well-established in numerous states, many of them have already been relied on to reduce air pollutants, including CO_2, from fossil fuel-fired EGUs and, as noted, they may be undertaken by the affected EGUs or, in general, required by the states.

For the alternative approach for the BSER, the EPA is identifying the "system of emission reduction" as including, in addition to building block 1, the reduction of affected fossil fuel-fired EGUs' mass emissions achievable through reductions in generation of specified amounts from those EGUs. Under this approach, the measures in building blocks 2, 3, and 4 would not be components of the system of emission reduction, but instead would serve as bases for quantifying the reduction in emissions resulting from the reduction in generation at affected EGUs. In light of the available sources of replacement generation through the measures in the building blocks, this approach also meets the criteria for being the "best" system because of, among other things, the emission reductions it would achieve, its reasonable cost, its promotion of technological development, as well as the fact that under this approach, the reliability of the electricity system would be maintained. The approach of reduced generation is also "adequately demonstrated" because of the ability of affected EGUs to adjust their own generation, the authority of the state to impose requirements, and the fact that other entities that operate in the various types of markets in the states can be expected to respond to the reduction in generation from the fossil-fuel fired EGUs by undertaking the measures in the building blocks or other actions that would assure reliability.

After determining BSER, the EPA is authorized under the implementing regulations, as an integral component to setting emission guidelines, to apply the BSER and determine the resulting emission limitation. The EPA is proposing to apply the BSER to the affected EGUs on a statewide basis. In this rulemaking, the EPA terms the resulting emission limitation the state goal. The EPA is formulating each state goal as an average emissions rate. The state goals form the EPA's emission guidelines.

With the promulgation of the emission guidelines, each state must develop a plan to achieve an emission performance level that corresponds to the state goal. The state plans must establish standards of performance for the affected EGUs and include measures that implement and enforce those standards. Based on requests from states and other stakeholders, the EPA is proposing that states be authorized to submit state plans that do not impose legal responsibility on the affected EGUs for the entirety of the emission performance level, but instead, by adopting what this preamble refers to as a "portfolio approach," impose requirements on other affected entities -- e.g., renewable energy and demand-side energy efficiency measures -- that would reduce CO_2 emissions from the affected EGUs. (In the preamble and the regulatory text for this proposed rulemaking, we refer to the affected EGUs and other entities with obligations under the state plan as "affected entities.") As noted in the preamble for this rulemaking, a possible basis for this approach is that those requirements on affected entities other than affected EGUs may be authorized as standards of performance or implementing measures. In the preamble, the EPA proposes that this is an appropriate flexibility and solicits comment, but also solicits comment on whether state plans must impose all of the legal responsibility for achieving the required emission performance level on the affected EGUs.

It should be noted that an important aspect of the BSER for affected EGUs is that the EPA is proposing to apply it on a statewide basis. The statewide approach also underlies the required emission performance level, which, as noted, is based on the application of the BSER to a state's affected EGUs, and which the suite of measures in the state plan, including the emission standards for the affected EGUs, must achieve overall. The state has flexibility in assigning the emission performance obligations to its affected EGUs, in the form of standards of performance -- and, for the portfolio approach, in imposing requirements on other affected entities -- as long as, again, the required emission performance level is met.

This state-wide approach both harnesses the efficiencies of emission reduction opportunities in the interconnected electricity system and is fully consistent with the principles of federalism that underlie the Clean Air Act generally and CAA section 111(d) particularly. That is, this provision achieves the emission performance requirements through the vehicle of a state plan, and provides each state significant flexibility to take local circumstances and state policy goals into account in determining how to reduce emissions from its affected sources, as long as the plan meets minimum federal requirements.

This state-wide approach, and the standards of performance for the affected EGUs that the states will establish through the state-plan process, are consistent with the applicable CAA section 111 provisions.

The preamble further notes that even if the state plan imposes all of the obligations to achieve the required emission performance level on the affected EGUs, the state plan could nevertheless include requirements on other affected entities in order to facilitate the reduced utilization of, and CO_2 emissions from, the affected EGUs – and the practical effect for the EGUs would be the same as under the proposed portfolio approach. The preamble solicits comment on other issues concerning state plans, including whether a state may include in its plan a mechanism to achieve a specified portion of the required emission performance level on behalf of the affected EGUs, and thereby limit the obligations of the affected EGUs.

The EPA emphasizes that in developing the state plans, the states have substantial discretion in designing the standards of performance, as long as the plans reduce emissions from the affected sources to achieve the required emission performance level. Moreover, the states may require sources to implement specific measures that the EPA does not identify as part of the BSER, and may include other approaches such as, for example, emission trading programs. By the same token, states may allow sources, in

19

complying with their applicable standards of performance, to rely on any measures that will reduce their CO_2 emissions, regardless of whether the EPA identifies those measures as part of BSER, as long as, again, the state plan achieves the requisite level of emissions reduction from the affected entities.

In this rulemaking, the EPA proposes reasonable deadlines for state plan submission and the EPA's action. The proposed deadline for the EPA's action on state play submittals varies from that in the implementing regulations, and the EPA is proposing to revise that provision in the regulations accordingly. Under CAA section 111(d)(2), the state plans must be "satisfactory" for the EPA to approve them, and in this rulemaking, the EPA is proposing the criteria that the state plans must meet under that requirement.

III. Authority to regulate CO_2 from EGUs

CAA section 111 authorizes EPA to regulate CO_2 emissions. The Supreme Court has held that greenhouse gases (including CO_2) are an "air pollutant" under the CAA. *Massachusetts. v. EPA*.[21] Furthermore, the U.S. Supreme Court's holding in *American Electric Power Co. v. Connecticut*, 131 S. Ct. 2527, 2537-38 (2011), that "the

---

[21] 549 U.S. 497 (2007).

Clean Air Act and the EPA actions it authorizes displace any federal common law right to seek abatement of carbon-dioxide emissions from fossil fuel-fired power plants" was premised on the Court's understanding that section 111, including section 111(d), applies to carbon dioxide emissions from those sources.

The fact that EPA has regulated EGU emissions of mercury and other hazardous air pollutants under CAA section 112 does not deprive EPA of the authority to regulate CO_2 emissions from EGUs under CAA section 111(d) under the Agency's established construction of the ambiguous provisions in CAA section 111(d)(1)(A)(i) that identify the air pollutants subject to CAA section 111(d). The ambiguities stem from apparent drafting errors that occurred during enactment of the 1990 CAA Amendments, which revised section 111(d). The confusion arises because two different amendments to section 111(d) were enacted in the 1990 CAA Amendments - one in title I of the bill, the other in title III of the bill (both amendments were to be codified in section 111(d)). The confusion is exacerbated because the U.S. Code does not accurately reflect what was enacted - it presents only one of the two amendments. However, the enacted law signed by the President (as recorded in the U.S. Statutes at Large), not the U.S. Code, is controlling.

As presented in the U.S. Code, section 111(d)(1)(A) requires states to submit standards of performance for existing sources "for any air pollutant (i) [1] for which air quality criteria have not been issued or which is not included on a list published under [CAA section 108(a)] or [2] *emitted from a source category which is regulated under [section 112]*." (Emphasis added.) This provision has two components that exclude from section 111(d) two types of air pollutants. The first component, which we call the NAAQS Exclusion, excludes NAAQS pollutants. The second component, which we call the Section 112 Exclusion, presents the ambiguities. As presented in the U.S. Code, the Section 112 Exclusion appears by its terms to preclude from section 111(d) any pollutant if it is emitted from a source category that is regulated under section 112. The U.S. Code version of 111(d) can be read to provide that the provision would not cover GHGs because GHGs are emitted from EGUs and EGUs are a source category regulated under section 112.[22]

_____

[22] By the same token, GHGs are emitted by many other source categories, such as refineries, that are regulated under section 112. Indeed, the text as presented in the U.S. Code could be read to exclude virtually every pollutant

22

The text of section 111(d) as presented in the U.S. Code, however, does not accurately reproduce the Section 112 Exclusion as enacted in the 1990 CAA Amendments. The correct statement of the Section 112 Exclusion – the one that was enacted by Congress and signed by the President, and which therefore is controlling – is found in the U.S. Statutes at Large. This text incorporates two versions of the Section 112 Exclusion, one passed by the U.S. House of Representatives and one passed by the U.S. Senate. The two versions were never reconciled, and both were enacted as part of the 1990 CAA Amendments. The two versions conflict with each other and thus render the Section 112 Exclusion ambiguous. Under these circumstances, the EPA may reasonably construe the Section 112 Exclusion to authorize the regulation of GHGs under section 111(d).

To understand the different amendments by the House and Senate, one must start with section 111(d)(1) as it read before the 1990 CAA Amendments:

> The Administrator shall prescribe regulations which shall establish a procedure similar to that provided by section 7410 of this title under which each State shall submit to the Administrator a plan which (A) establishes

---

from regulation under Section 111(d), because it would be difficult to identify any pollutant that is not emitted from at least one source category that is regulated under 112. We do not need to address this ridiculous result, however, for the reasons discussed in the text above.

standards of performance for any existing source
for any air pollutant (i) for which air quality
criteria have not been issued or which is not
included on a list published under section
7408(a) *or 7412(b)(1)(A) of this title,* but (ii)
to which a standard of performance under this
section would apply if such existing source were
a new source. * * *

42 U.S.C.A. 7411(d)(1) (West 1977); Public Law 95-95

(emphasis added). In this version, the Section 112

Exclusion, by its terms, applied to section 112 pollutants,

and not to categories of sources that emit those

pollutants. It should also be noted that in the 1990 CAA

Amendments, Congress amended section 112 to include a

statutory list of hazardous air pollutants for EPA to

regulate, instead of relying on EPA to develop its own

list.

The 1990 Senate bill amended revised section

111(d)(1)(A)(i) by striking the term to ''112(b)(1)(A)''

and inserting in its place the term ''112(b).'' Under this

amendment, the text would read as follows (with changes

shown in strikeout):

The Administrator shall prescribe regulations
which shall establish a procedure similar to that
Provided by section 7410 of this title under
which each State shall submit to the
Administrator a plan which (A) establishes
standards of performance for any existing source
for any air pollutant (i) for which air quality
criteria have not been issued or which is not

24

included on a list published under section
7408(a) or 7412(b) ~~(1)(A)~~ of this title, but (ii)
to which a standard of performance under this
section would apply if such existing source were
a new source.

The 1990 House bill amended section 111(d)(1)(A)(i) of
the 1977 CAA by striking the phrase ''or 112(b)(1)(A)'',
and inserting in its place the phrase ''or emitted from a
source category which is regulated under section 112.''
Under this amendment, the text would read as follows (with
changes shown in underline and strikeout):

> The Administrator shall prescribe regulations
> which shall establish a procedure similar to that
> Provided by section 7410 of this title under
> which each State shall submit to the
> Administrator a plan which (A) establishes
> standards of performance for any existing source
> for any air pollutant (i) for which air quality
> criteria have not been issued or which is not
> included on a list published under section
> 7408(a) or ~~7412(b)(1)(A)~~ <u>emitted from a source</u>
> <u>category which is regulated under section 7412</u> of
> this title, but (ii) to which a standard of
> performance under this section would apply if
> such existing source were a new source.

The House-Senate Conference Committee did not
reconcile these two conflicting amendments, and both were
included in the 1990 CAA Amendments as reported by the
Conference Committee, approved by both the House and the
Senate, and signed by the President. As presented in the
Statutes at Large, the Section 112 Exclusion is therefore
ambiguous.

The EPA discussed these different amendments in the
preamble to "Revision of December 2000 Regulatory Finding
on the Emissions of Hazardous Air Pollutants From Electric
Utility Steam Generating Units and the Removal of Coal- and
Oil-Fired Electric Utility Steam Generating Units From the
Section 112(c) List," 70 FR 15994, 16029-32 (March 29,
2005). There, the EPA concluded that the Section 112
Exclusion could be read as follows: Where a source category
is regulated under section 112, a section 111(d) standard
of performance cannot be established to address any HAP
listed under section 112(b) that may be emitted from that
particular source category. The EPA explained that this
approach reasonably interprets the Section 112 Exclusion to
give some effect to both amendments. The EPA emphasized
that it is not reasonable to give full effect to the House
language because a literal reading of that language would
mean that the EPA could not regulate any air pollutant from
a source category regulated under section 112, a result
that would be inconsistent with (i) Congress' desire in the
1990 CAA Amendments to require the EPA to regulate more
substances, and not to eliminate the EPA's ability to
regulate large categories of air pollutants, and (ii) the
fact that the EPA has historically regulated non-hazardous
air pollutants under section 111(d), even where those air

pollutants were emitted from a source category actually regulated under section 112.   *See* 70 FR 16031-32. The EPA continues to view this interpretation of the Section 112 exclusion as reasonable, for the reasons just stated.

Applying this interpretation of the Section 112 Exclusion to this rule, we conclude that section 111(d) authorizes the EPA to establish section 111(d) guidelines for GHG emissions from EGUs. Although EGUs are a source category that is regulated under CAA section 112, GHGs are not a HAP regulated under section 112. Therefore, the Section 112 exclusion in section 111(d) does not apply to GHGs, and 111(d) does not preclude the EPA from establishing guidelines covering GHGs from EGUs.

IV. Rational basis, endangerment finding

In response to the January 2014 Proposal for standards of performance for GHGs emissions from newly constructed fossil fuel-fired EGUs,[23] some stakeholders raised concerns that the EPA could not promulgate those standards without first issuing a finding that GHGs from those sources cause or contribute significantly to air pollution which may reasonably be anticipated to endanger public health or welfare, under CAA section 111(b)(1)(A). In that proposal, the EPA stated that it is rational to regulate GHGs from

---

[23] 79 Fed. Reg. 1,430 (Jan. 8, 2014).

fossil fuel-fired EGUs because the EPA has previously found that GHG emissions endanger public health and welfare, and because the electric generating industry emits a significant amount of GHGs.   The EPA added that CAA section 111 does not require that EPA issue a formal endangerment finding, and that even if section 111 did require such a finding, the EPA's rational basis would qualify as one.[24] The EPA is taking the same position in the section 111(b) rulemaking proposal to establish standards of performance for GHG emissions from modified and reconstructed fossil fuel-fired EGUs.

The EPA will finalize either or both of the January 2014 Proposal and the rulemaking for modified and reconstructed EGUs by the time that it finalizes this proposed rulemaking. In that event, the EPA would not be required to further address the rational basis or endangerment finding in this rulemaking. In any event, these questions are properly addressed and resolved in the context of the parallel rulemakings under section 111(b), not in this rulemaking. Thus, the EPA is not seeking comment in the preamble to this proposal on any issues related to a rational basis or endangerment finding.

---

[24] *See* 79 FR at 1,452/3 - 1,456/1.

V. Authority for EPA to determine BSER and emission guidelines

In this section we describe the authority, as set out in the EPA's implementing regulations under CAA section 111(d), for the EPA to determine the "best system of emission reduction … adequately demonstrated" and the amount of required emission reduction that is based on the BSER. We also describe how, in this rulemaking, the EPA proposes to apply the BSER to each state, and on that basis, to determine the amount of emission limitation achievable by each state, which we refer to as the state goal. The state goal is the "emissions guideline" that the implementing regulations require the EPA to promulgate.

CAA section 111(d) directs the EPA to –

> prescribe regulations which shall establish a
> procedure similar to that provided by [CAA
> section 110] under which each State shall submit
> to the Administrator a plan which (A) establishes
> standards of performance for any existing source
> for [certain air pollutants] … and (B) provides
> for the implementation and enforcement of such
> standards of performance.

As noted above, the EPA promulgated the implementing regulations for section 111(d) in 1975, and has revised parts of them since. The regulations set out a multi-step process for the development and approval of state plans, and assign responsibility for the various steps in the process to the EPA or the states. The EPA has followed these regulations in promulgating previous rulemakings under section 111(d).[25] In the present rulemaking, EPA continues to follow them, except that EPA is establishing a different deadline for submission of state plans than what the regulations would otherwise require.[26]

Under the implementing regulations, at the same time or after the EPA proposes and then finalizes standards of performance for sources in a source category under section 111(b), the EPA must propose and then finalize a "guideline document" with information pertinent to state plans under section 111(d):

> Concurrently upon or after proposal of standards
> of performance for the control of a designated
> pollutant from affected facilities, the
> Administrator will publish a draft guideline
> document containing information pertinent to
> control of the designated pollutant form [sic:
> from] designated facilities. Notice of the
> availability of the draft guideline document will
> be published in the Federal Register and public
> comments on its contents will be invited. After
> consideration of public comments and upon or
> after promulgation of standards of performance
> for control of a designated pollutant from
> affected facilities, a final guideline document
> will be published and notice of its availability
> will be published in the Federal Register."[27]

The regulations go on to describe the contents of the "guideline document" as including, among other things, an

---

[25] These rulemakings are cited above.
[26] The EPA is not re-opening these regulations, although it is revising the deadline for EPA action on state plans.
[27] 40 CFR 60.22(a).

"emission guideline" that incorporates the "best system of

emission reduction … adequately demonstrated":

> Guideline documents published under this section
> will provide information for the development of
> State plans, such as: * * * *

> (5) An emission guideline that reflects the
> application of the best system of emission
> reduction (considering the cost of such
> reduction) that has been adequately demonstrated
> for designated facilities, and the time within
> which compliance with emission standards of
> equivalent stringency can be achieved.* * * *

> (6) Such other available information as the
> Administrator determines may contribute to the
> formulation of State plans.[28]

The implementing regulations define the "emission

guideline" as –

> A guideline set forth in subpart C of this part,
> or in a final guideline document published under
> section 60.22(a) which reflects the degree of
> emission reduction achievable through the
> application of the best system of emission
> reduction which (taking into account the cost of
> such reduction) the Administrator has determined
> has been adequately demonstrated for designated
> facilities.[29]

In addition, the implementing regulations mandate that for

air pollutants that adversely affect public health, the

"emission guidelines" must be proposed and finalized with

the draft and final guideline document:

> [For air pollutants that have been demonstrated
> to adversely affect public health], the emission

---

[28] *Id.* at 60.22(b).
[29] 40 CFR 60.21(e).

guidelines and compliance times referred to in
paragraph (b)(5) of this section will be proposed
for comment upon publication of the draft
guideline document, and after consideration of
comments will be promulgated in subpart C of this
part with such modifications as may be
appropriate.[30]

With this proposed rulemaking, the EPA is complying

with these regulatory provisions. This proposed rulemaking

follows the proposal of standards of performance for newly

constructed affected sources in the January 2014 Proposal,

and is concurrent with the proposal of standards of

performance for modified and reconstructed affected

sources. This proposed rulemaking – including the preamble

and the supporting documents -- comprise the "draft

guideline document." The documents contain the "information

for the development of State plans" described in the

regulations. This information includes descriptions as well

as technical and economic evaluations of the four building

blocks. This information also includes the EPA's

application of the BSER to each state, and the EPA's

calculation of the resulting proposed state goals. These

state goals comprise the proposed "emission guidelines."

In addition, the preamble and supporting documents propose

the "time within which compliance with emission standards

_____

[30] *Id.* at 60.22(c).

of equivalent stringency can be achieved," which are the periods of 2020-2029 for interim compliance, and the subsequent period for final compliance, and provide other information.

VI. Best system of emission reduction adequately demonstrated and standards of performance

In this section we discuss our interpretation of the CAA section 111(d)(1) and (a)(1) provisions that require the state plans to establish, for "any existing source," "standards of performance," and that define the latter term to mean, in general, emission standards that "reflect the degree of emission limitation achievable through the application" of the "best system of emission reduction … adequately demonstrated" (BSER).

In subsection A of this section, we explain these section 111(d)(1) and 111(a)(1) provisions and summarize key parts of the applicable case law.

In subsection B, we describe our proposed two alternative determinations for the BSER. We note that each alternative includes two main components. One component, for each alternative, is efficiency improvements that coal-fired power plants can make to their operations and equipment (which we call building block 1). For the first type of BSER, the remaining component is, in general, increased zero- or low-emitting generation in specified amounts (building blocks 2 and 3), and increased demand-side energy efficiency in specified amounts (building block 4), all of which have the effect of displacing generation from the higher-emitting affected sources. For the alternative type of BSER, the remaining component is reduced generation from higher-emitting affected sources in specified amounts, which is the amount that can be replaced by, in general, increased zero- or low-emitting generation and eliminated by increased demand-side energy efficiency. After we explain these alternatives, we go on to discuss why each alternative is a "system of emission reduction," and why we propose to determine that each system is the "best" that is "adequately demonstrated."

In subsection C, we discuss our interpretation of the requirement that each state must develop a plan that establishes for "any existing source" "standards of performance," that is, emission standards that "reflect the degree of emission limitation achievable through the application of the [BSER]." We explain that once the EPA determines the BSER, it undertakes "the application of the [BSER]" to each state's set of sources on a state-wide basis, and thereby determines the "emission limitation achievable," which we term the state goal, and which in turn becomes the required emission performance level that the state plan must achieve. The state must then develop its plan by identifying emission standards for its affected EGUs -- and, in the case of a state that adopts the portfolio approach, by identifying other obligations on other affected entities -- that in total, achieve the required emission performance level. Through this process, the state plan may meet the requirements of sections 111(d)(1) and (a)(1) to "establish[] standards of performance for any existing source" because it imposes, on each of its affected sources, emission standards that "reflect [that is, embody or represent] the degree of [that is, the part of] emission limitation achievable through the application of the best system of emission reduction … adequately demonstrated" [that is, the state's required emission performance level].

A. CAA requirements for standards of performance and BSER

In this subsection, we explain the relevant provisions of sections 111(d)(1) and 111(a)(1) and summarize key parts of the applicable case law.

The EPA's explanation for this BSER proposal begins with the key statutory provisions in CAA sections 111(d)(1) and 111(a)(1). Section 111(d)(1) requires that a state plan "(A) establish[] … standards of performance for any existing source" and "(B) provide[] for the implementation and enforcement of such standards." Section 111(a)(1) defines a "standard of performance" as --

> a standard for emissions of air pollutants which reflects the degree of emission limitation achievable through the application of the best system of emission reduction which (taking into account the cost of achieving such reduction and any nonair quality health and environmental impact and energy requirements) the Administrator determines has been adequately demonstrated.

Several points should be made about the BSER. By its terms, it is a "system of emission reduction" that is both the "best" and "adequately demonstrated." The CAA does not define the term, "system," and as a result, that term should be given its ordinary, everyday meaning: "a set of things working together as parts of a mechanism or interconnecting network; a complex whole."[31] In addition, the U.S. Court of Appeals for the D.C. Circuit (D.C. Circuit or Court) has handed down case law over a 40-year period that interprets the requirements that the "system of

---

[31] *Oxford Dictionary of English* (3rd ed.) (published 2010, online version 2013) http://www.oxfordreference.com.mutex.gmu.edu/view/10.1093/acref/9780199571123.001.0001/acref-9780199571123

emission reduction be the "best" and be "adequately demonstrated."[32] Under this case law, the criteria for the EPA to use in determining whether the system is the "best" include the following key considerations, among others:

- The system of emission reduction must be technically feasible.[33]

- The EPA must consider the amount of emissions reductions that the system would generate.

- The costs of the system must be reasonable. The EPA may consider the costs on the source level, the industry-wide level, and, at least in the case of the power sector, on the national level in terms of the overall costs of electricity and the impact on the national economy over time.[34]

---

[32] *Portland Cement Ass'n v. Ruckelshaus*, 486 F.2d 375 (D.C. Cir. 1973), *cert. denied,* 417 U.S. 921 (1974); *Essex Chemical Corp. v. Ruckelshaus*, 486 F.2d 427, (D.C. Cir. 1973), *cert. denied, Appalachian Power Co. v. EPA*, 416 U.S. 969 (1974); *Sierra Club v. Costle*, 657 F.2d 298 (D.C. Cir. 1981); *Portland Cement Ass'n v. EPA*, 665 F.3d 177 (D.C. Cir. 2011).

[33] The case law may be read to treat technical feasibility as the measure for whether the standard of performance is "achievable," Essex Chemical Corp. v. Ruckelshaus, 486 F.2d at 427, not as a criterion for whether the system of emission reduction is the "best system of emission reduction … adequately demonstrated." However, for present purposes, we refer to technical feasibility as another of the criteria for the BSER.

[34] *See Sierra Club v. Costle,* 657 F.2d 298, 330-31, 337-39 (D.C. Cir. 1981). As discussed in the January 2014 Proposal, the D.C. Circuit's case law formulates the cost consideration in various ways: the costs must not be "exorbitant[]", *Essex Chemical Corp. v. Ruckelshaus*, 486 F.2d 427, 433 (D.C. Cir. 1973) *cert. denied, Appalachian Power Co. v. EPA,* 416 U.S. 969 (1974), *see Lignite Energy Council v. EPA*, 198 F.3d 930, 933 (D.C. Cir. 1999); "greater than the industry could bear and survive," *Portland Cement Ass'n v. EPA*, 513 F.2d 506, 508 (D.C. Cir. 1975); or "excessive" or "unreasonable." *Sierra Club v.*

- The EPA must also consider that CAA section 111 is designed to promote the development and implementation of technology.[35]

- The EPA must also consider energy impacts, and, as with costs, may consider them both on the source level and on the nationwide structure of the power sector over time.

Importantly, the EPA has discretion to weigh these various considerations, may determine that some merit greater weight than others, and may vary the weighting, depending on the source category.

In determining whether a system is "adequately demonstrated," the EPA is to look forward toward what may fairly be projected for the regulatory future, rather than determining what is available now. In the D.C. Circuit's first decision under section 111, *Portland Cement Ass'n v. Ruckelshaus*, 486 F.2d 375, 391 (D.C. Cir. 1973), the Court explained:

> Section 111 looks toward what may fairly be projected for the regulatory future, rather than the state of the art at the present . . . . The Senate Report made clear that it did not intend that the technology "must be in actual routine

---

Costle, 657 F.2d 298, 343 (D.C. Cir. 1981). In the January 2014 Proposal, EPA stated that "these various formulations of the cost standard ... are synonymous," and, for convenience, EPA used "reasonableness" as the formulation. EPA takes the same approach in this rulemaking.
[35] *See* 79 Fed. Reg. at 1,465/1-2 (discussing case law and legislative history that includes technological development as a consideration in the determination of BSER, including *Sierra Club v. Costle*, 657 F.2d 298 (D.C. Cir. 1981)).

use somewhere." . . . The Administrator may make a projection based on existing technology, that that projection is subject to the restraints of reasonableness and cannot be based on "crystal ball" inquiry. . . . [T]he question of availability is partially dependent on "lead time", the time in which the technology will have to be available.[36]

The forward looking nature of determining whether a system is adequately demonstrated is particularly relevant for this proposal given the lengthy period for implementing state plans that the EPA is proposing. The EPA discussed the CAA requirements and Court interpretations of the BSER at length in the January 2014 Proposal,[37] and incorporates by reference that discussion in this rulemaking.

It should be noted that the EPA may identify as the best system of emission reduction adequately demonstrated as the BSER a system that would form the basis for emission standards that could be achieved by some, but not necessarily all, of the existing sources in the source category. This approach is consistent with the technology-forcing purposes of section 111, as well as the fact that under section 111(d)(1), the state retains authority, "in applying a standard of performance to any particular source

---

[36] *Portland Cement Ass'n v. Ruckelshaus*, 486 F.2d 375, 391–92 (D.C. Cir. 1973) (citations omitted).
[37] *See* 79 Fed. Reg. at 1,462/1 – 1,467/3.

... to take into consideration, among other factors, the remaining useful life of the ... source...."[38]

B. Best system of emission reduction adequately demonstrated

In this subsection, we describe our two alternative proposed determinations for the BSER and explain why each

---

[38] The EPA discussed this issue in connection with new sources in the recently proposed NSPS for CO_2 emissions from fossil fuel-fired EGUs, 79 Fed. Reg. 1430, 1466/3 (Jan. 8, 2014). With respect to existing sources, a commentator has stated:

> There is no statutory provision or direct precedent under § 111(d) requiring EPA to demonstrate that emission limits are achievable by every source subject to an [standard of performance for existing sources]. Moreover, since the trigger for implementing § 111(d) is an NSPS under § 111(b), Congress arguably contemplated that, once EPA has identified BSER for new plants, it should raise the performance of the existing fleet with the goal of approaching new source levels at existing plants. In this reading, 111(d) would have a technology-forcing thrust, tempered by the performance and cost constraints at existing plants but nonetheless raising the bar significantly for the existing fleet. From this perspective, EPA could argue that "adequately demonstrated" means achievable at a reasonable cost by the more modern, better performing coal and gas units, not by all plants [citing Reinforcing this approach is the fact that cost is not determinative in defining a "standard of performance" under § 111(a) but only must be "taken into account" ....

Sussman, R., "Power Plant Regulation Under the Clean Air Act: A Breakthrough Moment for U.S. Climate Policy?," *Virginia Environment Law Journal*, 32:97 (2014), at 123 (citations omitted).

is a "system of emission reduction," and why each system is the "best" that is "adequately demonstrated."

1. Introduction and overview

The EPA's BSER proposal in this rulemaking recognizes, and is based in part on, the interconnected nature of the electrical generating system, which, among other things, means that generation at one EGU can substitute for generation at another. The importance of the interconnected nature of the grid in facilitating CO_2 emissions reductions is evident in the long history of reliance on it to provide least-cost dispatch, the more recent history of implementing air pollutant emissions reductions, and the still more recent history of implementing CO_2 emissions reductions at the company, state, and regional level.

In this rulemaking, the EPA proposes to determine the "best system of emission reduction … adequately demonstrated" on a state-by-state basis. Moreover, the EPA proposes to determine the BSER based on four "building blocks," some of which rely on the interconnected nature of the electricity generating grid:

Building block 1: Reducing the carbon intensity of generation at individual affected EGUs through heat rate improvements.

41

Building block 2: Reducing emissions from the most carbon-intensive affected EGUs in the amount that results from substituting generation at those EGUs with generation from less carbon-intensive affected EGUs (including NGCC units under construction).

Building block 3: Reducing emissions from affected EGUs in the amount that results from substituting generation at those EGUs with expanded low- or zero-carbon generation.

Building block 4: Reducing emissions from affected EGUs in the amount that results from the use of demand-side energy efficiency that reduces the amount of generation required.

As discussed in the preamble, with these building blocks in mind, we are proposing two alternatives for the BSER, each of which is based on methods for reducing fossil fuel-fired EGUs' air pollutants that states and sources have already implemented. The first approach is that the BSER is the combination of building blocks 1 through 4. Building block 1 is a set of operational improvements and equipment upgrades that the affected sources may undertake to improve their efficiency and reduce their emissions rate. Building blocks 2, 3 and 4 are sets of measures that, in general, increase zero- or low-emitting generation in specified amounts and increased demand-side energy efficiency in specified amounts, all of which, due to the interconnected nature of the grid, result in drawing utilization away from higher-emitting fossil fuel-fired

EGUs, thereby lowering those EGUs' emissions. The second approach is that the BSER is building block 1 (heat rate improvements) combined with reduced generation from fossil fuel-fired EGUs in the amount, calculated on a statewide basis, that can be replaced by, in general, increased zero- or low-emitting generation and avoided by increased demand-side energy efficiency. The EPA proposes that each of these alternatives may be considered to be a "system of emission reduction," and that each meets the criteria, set out in CAA section 111(a)(1) and the case law, to qualify as the "best" system that is "adequately demonstrated."

2. Background: Interconnected nature of the electricity system

Central to our BSER determination is the fact that the nation's electricity needs are being met, and have for many decades been met, through a grid formed by a network connecting groups of EGUs with each other and, ultimately, with the end-users of electricity. We discuss this nature of the electricity system at length in the preamble and, for convenience, summarize that discussion here.

Through the interconnected grid, fungible products - electricity and electricity services - are produced and delivered by a diverse group of EGUs operating in a coordinated fashion in response to end-users' demand for electricity. Because the electricity grid operates through the interconnection of multiple EGUs and favors least-cost generation, owners and operators of generators have been able to assure the stability of electricity generation and the reliable delivery of electricity to users at least cost (subject to certain reliability, environmental and other constraints). The fact that generation at one EGU can be substituted for generation at another allows operators to utilize their least-cost assets first, and hold their higher-cost assets in reserve, thereby assuring that the system achieves the objectives of providing reliable and least-cost electricity service.

In recent years, the ability to shift between different generation assets on the grid has also facilitated the achievement of environmental objectives, including the reduction of emissions of nitrogen oxides, sulfur dioxide, and particulate matter -- which, among them, worsen acid deposition and jeopardize the attainment and maintenance of national ambient air quality standards - as well as hazardous air pollutants. Regulation of those air pollutants tends to increase the relative cost of electricity from higher-emitting generation assets. Because EGU operators have the ability to use the grid as an interchange for shifting levels of generation among several facilities, the higher costs of higher-emitting assets must be considered, along with fuel costs and other marginal costs, in determining the extent to which those assets are utilized. The amount of their utilization affects the amount of their emissions.

Most recently, states and companies seeking specifically to achieve CO_2 emissions reduction objectives have also relied on the shifting of generation between and among EGUs to achieve those emissions reduction objectives. In fact, as the preamble notes, there are many cases in which companies have reduced emissions through shifting generation away from higher emitting units to lower- or zero-emitting units, or through reducing overall electric demand through demand-side energy efficiency measures. In some cases, this has occurred in response to goals set at the company level: some companies have established a single, company-wide emission target, and then have used combinations of strategies such as fuel switching, increased renewable or nuclear generation, and increased energy efficiency, to achieve those goals. In other cases, this has occurred in response to goals set at the state level: for example, California enacted its Global Warming Solutions Act in 2006 (AB 32), requiring the state to reduce its GHG emissions to 1990 levels by 2020 and 80 percent below 1990 levels by 2050,[39] through a suite of mechanisms that include energy efficiency programs, renewable energy programs and an economy-wide cap and trade program, along with other programs.[40] Similarly, nine northeast and mid-Atlantic states participate in the Regional Greenhouse Gas Initiative (RGGI), a market-based emissions budget trading program that sets an aggregate limit on CO_2 from fossil fuel fired power plants in the participating states. These examples demonstrate that it is appropriate to base the BSER at least in part on the combination of measures in building blocks 2, 3, and 4, and that this component of such a system is adequately demonstrated.

In all of these instances, companies' choices and policies implemented by states may impact decisions about dispatching of lower instead of higher emitting generating units both as part of the short term dispatch process and as part of longer term business planning processes. The proposed emission guidelines, including the temporal

---

[39] State of California Global Warming Solutions Act of 2006, Assembly Bill 32, Chapter http://www.leginfo.ca.gov/pub/05-06/bill/asm/ab_0001-0050/ab_32_bill_20060927_chaptered.pdf
[40] *See* Cal. Air Res. Bd., Climate Change Scoping Plan 31-32, 41-46 (2008), *available at:* http://www.arb.ca.gov/cc/scopingplan/document/adopted_scoping_plan.pdf.

flexibility that the guidelines incorporate, allow states and EGUs to implement a variety of mechanisms that can reduce emissions both as part of those shorter term dispatch decisions and as part of longer term business planning processes.

In this rulemaking, the EPA is building on these company, state, and regional approaches by continuing to rely on the interconnected nature of the grid to achieve, on a nationwide basis, the important objective of significant amounts of CO_2 reductions from fossil fuel-fired EGUs. The EPA is doing so by proposing that the BSER should be based on a combination of the implementation of heat improvement measures for fossil fuel-fired steam generating units (building block 1) to reduce their emissions, as well as the implementation of other measures that are associated with reduced emissions from those EGUs. The latter include substituting generation at higher emitting sources with increasing generation at less carbon-intensive EGUs, using expanded amounts of low- or zero-carbon generating capacity connected to the electric grid, and using electricity more efficiently to reduce the total demand for electricity (building blocks 2, 3 and 4, respectively).

In determining the BSER, it is significant that CO_2 is a global pollutant, and therefore the location of the

47

emissions (or emission reductions) does not affect the
impact on climate change of an amount of emissions
generated at any given source in any one location.[41] The
fact that CO_2 becomes well-mixed in the atmosphere means
that CO_2 emissions may be reduced anywhere within the
electricity grid and still achieve the intended climate
benefits. This allows the EPA to determine that a system
is the "best" system based on the total emission reductions
the system would achieve, rather than basing the
determination on the emission reductions achieved at each
individual affected source.

3. First Approach to the BSER: Building blocks 1, 2, 3, and 4 in combination

Under the EPA's first approach to determining the
BSER, the EPA is proposing that the BSER is the combination
of building blocks 1 through 4. As discussed in the

---

[41] By analogy, because the problem of acid deposition is
caused by EGU emissions of nitrogen oxide and sulfur
dioxide over a wide geographic area, Title IV of the Clean
Air Act established a national emissions trading program
that addresses that problem by reducing the total amount of
those emissions, but without regard to the particular
location of those emissions (or emissions reductions). In
contrast, other air pollutants have adverse health and
welfare effects in the locality where they are emitted, and
as a result, geographic constraints on emissions trading
are necessary. See CAA section 173(c)(1) (limiting offsets
for air pollutants subject to new source review
requirements to emissions reductions from sources in
certain nonattainment areas).

preamble, under the EPA's proposed approach to determining the BSER, the measures in building block 1, which entail improvements in the efficiency of the affected EGUs' equipment or processes, meet the criteria to qualify as a part of the BSER. Further elaboration of this point here is not necessary. In addition, in the preamble, we explain why all four building blocks in combination meet the criteria to qualify as the BSER, and further elaboration of this latter point here is also not necessary. Instead, this discussion will focus on building blocks 2, 3, and 4.

Under this first approach to the BSER, the "best system of emission reduction … adequately demonstrated" also includes the measures in building blocks 2, 3 and 4 for the affected fossil fuel-fired steam generating boilers, and building blocks 3 and 4 for the fossil fuel-fired combustion turbines. In this section, we first explain why the measures in these building blocks are part of a "system of emission reduction," and then why that system is the "best" system that is "adequately demonstrated."

In brief, building blocks 2, 3, and 4 are part of a "system of emission reduction" because that phrase, in the context in which it is used in section 111 and by its terms, is broad enough to apply to the measures in the building blocks, in light of the integrated nature of the electricity grid. Through the integrated grid, the measures reduce overall demand for, and therefore utilization of, higher emitting, fossil fuel-fired EGUs, which, in turn, reduces CO_2 emissions from those EGUs. The measures in the building blocks are part of the "best" system that is "adequately demonstrated" because they meet the criteria in section 111(a)(1) and the case law for BSER and they are well-established.

a. "[S]ystem of emission reduction"

The EPA's proposal that in this rulemaking, the "system of emission reduction" includes the measures in building blocks 2, 3, and 4 is grounded in the EPA's interpretation of the key CAA provisions: section 111(d)(1), which requires that each state plan "establish[] standards of performance for any existing source" for certain types of air pollutants; and section 111(a)(1), which defines a "standard of performance" as "a standard for emissions … which reflects the degree of emission limitation achievable through the application of the best system of emission reduction … adequately demonstrated." As explained next, the EPA's interpretation may be justified under either a *Chevron* step 1 or *Chevron* step 2 interpretation.

i. Chevron step 1 interpretation

The starting point for our analysis is the phrase, "system of emission reduction," which serves as the basis for the "standard for emissions." As noted above, the CAA does not define the term, "system," and as a result, that term should be given its ordinary, everyday meaning: "a set of things working together as parts of a mechanism or interconnecting network; a complex whole."[42] This definition is broad. It encompasses virtually any "set of things" that reduce emissions. Moreover, no other provisions in the definition of "standard of performance" include any other constraints on the type of "things" that may serve as the basis for the standard for emissions. The only constraints are the qualifiers "best" and "adequately demonstrated," but these do not constrain the type of "things" that could be a "system of emission reduction," only whether a particular "thing" qualifies as the "best" "system of emission reduction" that is "adequately demonstrated" (it must be, among other things, technically feasible and of reasonable cost). Thus, the "system of emission reduction" may include anything that reduces emissions, ranging from

---

[42] *Oxford Dictionary of English* (3rd ed.) (published 2010, online version 2013)
http://www.oxfordreference.com.mutex.gmu.edu/view/10.1093/acref/9780199571123.001.0001/acref-9780199571123

add-on controls applied to the affected sources'
smokestacks to control emissions, to measures that replace
production or generation at the affected sources and
thereby reduce emissions from those sources.

Moreover, the context in which "standard of
performance," which includes "system of emission
reduction," is found does not add additional constraints.
As noted above, section 111(d)(1) requires that state plans
establish "standards of performance for any existing
source," and in the preamble, we solicit comment on the
interpretation of that phrase. Among other things, we
solicit comment on whether the standards of performance
must apply directly to the affected sources and only to the
affected sources, in which case the affected sources would
bear the legal liability for the entire amount of emission
reduction requirements; or whether, instead, the standards
of performance may apply to other entities whose actions
would reduce generation, and thus emissions, from the
affected sources. Under either of those interpretations,
there is nothing in that phrase that limits the type of
"system of emission reduction" that, if it is the "best"
that is "adequately demonstrated," may furnish the basis
for the standards for emissions. That is, even if that
phrase -- "standards of performance for any existing

52

source" -- is interpreted to mean that the standards of performance must apply directly to, and only to, the affected sources, that application of the standards of performance does not limit the scope of the type of "system of emission reduction" that may serve as the basis for the standards for emissions. Any "system of emission reduction" that reduces the emissions of the affected sources may serve as the basis for the standards for emissions, as long as, again, it is the "best" that is "adequately demonstrated." For these reasons, the scope of the type of "system of emission reduction" that may be considered is broad.

Interpreting the "system of emission reduction" in this manner is also consistent with the scope of the state plans. Under section 111(d)(1), a state plan must "establish[] standards of performance" and "provide[] for the implementation and enforcement of such standards of performance." At the state's discretion, measures in building blocks 2, 3, and 4 may be included in state plans either through the portfolio approach or as measures that "provide[] for the implementation" of standards of performance that limit emissions from affected EGUs.

Based on these interpretations, for existing sources in the electric utility industry, we propose that the term

"system of emission reduction" is sufficiently broad to
include the measures in building blocks 2, 3, and 4 because
they are part of the interconnected electricity sector and
result in reduced utilization, and therefore reduced
emissions, from the higher emitting fossil fuel-fired power
plants. This proposed reading is clear as a matter of
Chevron step 1 because of the breadth of the term,
"system," in the context in which it is found.

ii. Chevron step 2 interpretation

Moreover, even if the term, "system of emission
reduction" is not considered to be clear as a matter of
Chevron step 1 to include the measures in building blocks
2, 3, and 4, then the EPA's interpretation of the term to
include those measures is valid as reasonable construction
under *Chevron* step 2. There are several reasons for
interpreting "system of emission reduction" in this way.[43]

(I). Legislative history of "standard of performance"

---

[43] *See EPA v. EME Homer City Generation, L.P.*, No. 12-1182,
slip op. at 22 (U.S. April 29, 2014) (after explaining why
the text of the CAA "did not answer" the largely technical
question of how EPA should allocate each state's
responsibility for the tangle of potentially "significant"
upwind-to-downwind air pollution contributions,
stating: "Under *Chevron*, we read Congress' silence as a
delegation of authority to EPA to select from among
reasonable options.")

First, the legislative history of the definition of "standard of performance," including the phrase "best system of emission reduction … adequately demonstrated," makes clear that the "system of emission reduction" is broader than a technological system. As enacted by Congress in the 1970 CAA Amendments, section 111(a)(1) defined the term "standard of performance" as, in relevant part --

> a standard for emissions of air pollutants which reflects the degree of emission limitation achievable through the application of the best system of emission reduction … adequately demonstrated.

In the 1977 CAA Amendments, Congress changed this definition to require that for new sources, the standard must, in relevant part, "reflect the degree of emission limitation … achievable through application of the best *technological* system of continuous emission reduction … adequately demonstrated;" and for existing sources, the standard must, in relevant, "reflect[] the degree of emission reduction achievable through the application of the best system of continuous emission reduction … adequately demonstrated…." (Emphasis added.)[44] In the 1990 CAA Amendments, Congress again changed this definition,

_____

[44] The 1977 CAA Amendments also revised section 111(a)(1) to require that the standards of performance for fossil fuel-fired sources require a percentage reduction in emissions (the "percentage reduction" requirement).

this time to reinstate the definition as found in the 1970 CAA Amendments (with some revisions not here relevant). That is, Congress repealed the requirements added in the 1977 CAA Amendments that the "system" be, in the case of new sources, "technological."

These amendments make clear that the "system[s] of emission reduction" upon which the section 111(d) standards of performance may be based are not limited to technological systems. Even when, in the 1977 CAA Amendments, Congress limited the systems that could provide the basis for the standards of performance for new sources to technological systems, Congress did not establish that limit on the systems for existing sources. Moreover, the 1977 House-Senate Conference Committee report stated that for existing sources, the standards of performance were to be based on the "best available means of emission control (not necessarily technological)…."[45]

(II). Pollution prevention

In addition, interpreting the term "system of emission reduction" broadly to include the building blocks is

---

[45] "Joint Explanatory Statement of the Committee of Conference," *reprinted in* Congressional Research Service, *A Legislative History of the Clean Air Act Amendments of 1977*, vol. 3 at 502, 509 (1978) (*1977 Legislative History*). The House Committee Report included the same statement. *See* H. Rep. 95-294 at 195, *reprinted in 1977 Legislative History,* vol. 4 at 2465, 2662.

consistent with a primary purpose of the CAA, which is encouraging pollution prevention, including assuring that states fulfill their role in developing pollution prevention measures. CAA section 101(c) states that "[a] primary goal of [the Clean Air Act] is to encourage or otherwise promote reasonable Federal, State, and local governmental actions, consistent with the provisions of this chapter, for pollution prevention." CAA section 101(b)(4) adds that one of "the purposes of [title I of the CAA, which includes section 111] are … (b) to encourage and assist the development and operation of regional air pollution prevention and control programs." Indeed, in the U.S. Code, in which the CAA is codified as chapter 85, the CAA is entitled, "Air Pollution Prevention and Control." CAA section 101(a)(3) describes "air pollution prevention" as "the reduction or elimination, through any measures, of the amount of pollutants produced or created at the source," and adds: "The Congress finds -- … (3) that air pollution prevention … and air pollution control at its source is the primary responsibility of States and local governments."

The measures in building blocks 2, 3, and 4 all qualify as types of "pollution prevention" because they are "measures" that "reduc[] or eliminate[e] … the amount of pollutants produced or created at the [fossil fuel-fired affected] source[s]." It is reasonable to interpret the section 111 provisions at issue in this rulemaking in light of these section 101 provisions, and this supports the reasonableness of interpreting the broad term found in section 111(a)(1), "system of emission reduction," to include the pollution prevention measures in building blocks 2, 3, and 4.

(III). Title IV

The breadth of the term, "system of emission reduction" is further confirmed by reference to certain provisions of CAA Title IV. In Title IV, Congress established the program that regulates fossil fuel-fired power plants to reduce their emissions of the precursers to acid deposition, including reducing sulfur dioxide (SO_2) emissions in two phases, and reducing nitrogen oxides (NOx) emissions. Congress enacted Title IV as part of the 1990 CAA Amendments, at the same time that Congress revised the definition of "standard of performance" to generally return it to its 1970-vintage reading. In certain respects, section 111 and Title IV are related because both apply to fossil-fuel fired EGUs, and Congress recognized the relationship in several Title IV provisions.[46]

One contrasting provision in Title IV is section 407(b)(2), which requires the EPA to base the NOx emission limits for certain types of boilers "on the degree of reduction achievable through the *retrofit* application of the best system of continuous emission reduction...;" and further requires the EPA to revise previously promulgated emission limits for certain types of boilers "to be more stringent if the [EPA] determines that *more effective low NOx burner technology* is available." (Emphasis added.) These narrower specifications for the basis of the Title IV emissions limits make clear that Congress knew how to constrain the basis for emission limits to the results of certain technology, and that its choice to base the section 111(d) standards of performance on a "system of emission

---

[46] *See, e.g.,* CAA section 402(8), 405(c)(2). In fact, in the 1990 CAA Amendments, Congress based its decision to repeal the percentage reduction requirements added in the 1977 CAA Amendments to the section 111(a)(1) definition of "standard of performance" for new fossil fuel-fired sources at least in part on the grounds that provisions of Title IV would cap SO_2 emissions from fossil-fuel fired EGUs, and, further, Congress conditioned that repeal on the continued applicability of the SO_2 cap, so that if the cap were eliminated, the repeal would, by operation of law, be eliminated. *See* Pub. L. 101-549 section 403(b), S. Rep. 101-228, at 338, *reprinted in* 1990 Legislative History 8338, 8678 (1990 Senate Committee Report).

reduction" indicates its intent to authorize a broader basis for those standards.

Other provisions in Title IV and their legislative history provide further support for interpreting the term, "system of emission reduction" to include building blocks 2, 3, and 4. In designing Title IV, Congress recognized the integrated nature of the electricity sector and how that integration could be harnessed to reduce air pollutant emissions; and, in fact, Congress included provisions to encourage re-dispatch to lower emitting sources, renewable energy, and demand-side energy efficiency, all of which are measures in those building blocks. Specifically, Congress added into the "purposes" provision of Title IV, the statements that in addition to the reducing the adverse effects of acid deposition –

> It is also the purpose of [Title IV] to encourage energy conservation, use of renewable and clean alternative technologies, and pollution prevention as a long-range strategy, consistent with the provisions of [Title IV], for reducing air pollution and other adverse impacts of energy production and use…."[47]

---

[47] CAA section 401(b). *See* H. Rep. 101-490 Part 1 at 369-70 (1990 House Comm. Rep.), *reprinted in* "A Legislative History of the Clean Air Act Amendments of 1990," *Congressional Research Service* (1993) (1990 Legislative History), vol. II, at 3021, 3393-94.

Congress recognized that the very structure of Title IV – which imposed a marketable trading system under which affected sources were required to have an allowance for each ton of SO_2 emitted and could buy and sell allowances on the open market -- encouraged such measures as demand-side energy efficiency and re-dispatch by lower-emitting sources. The 1990 Senate Committee Report explained:

> [T]he incentives created by the allowance market should stimulate innovations and the technologies and strategies used to reduce emissions…. [T]he allowance market should encourage sources to exploit *energy efficiency*, enhanced emission reduction or control technologies….; fuel-switching and *least-emissions dispatching* in order to maximize emission reductions."[48]

In addition, Congress incorporated into Title IV specific incentives to further encourage electric utilities (defined as entities that sell electricity[49]) to reduce their emissions through demand-side energy efficiency and renewable energy: Section 404(f)-(g) provided a special reserve of allowances to be allocated to electric utilities "for each ton of SO_2 emissions avoided by an electric utility … through the use of … energy conservation measures or … renewable energy." In fact, in adopting these provisions, Congress explicitly recognized the integrated

---

[48] 1990 Senate Committee Report at 316, *reprinted in* 1990 Legislative History, vol. V, at 8656 (emphasis added).
[49]   CAA section 404(f)(1)(C).

nature of the electricity sector: As one of the conditions

for eligibility for this special reserve of allowances, the

utility must "ha[ve] adopted and is implementing a least

cost energy conservation and electric power plan which

evaluates a range of resources, including new power

supplies, energy conservation, and renewable energy

resources, in order to meet expected future demand at the

lowest system cost."[50]

These CAA provisions and the accompanying statements

in the legislative history make clear that in enacting the

Title IV provisions to reduce SO_2 and NO_x emissions from

fossil fuel-fired EGUs, Congress viewed the electricity

sector as interconnected and considered re-dispatch to

lower emitting sources, renewable energy, and demand-side

energy efficiency as methods to reduce those emissions.

All this supports the reasonableness of the EPA's proposed

interpretation that the "system of emission reduction" that

serves as the basis for "standards of performance" for CO_2

emissions from fossil fuel-fired EGUs may include those

same measures, that is, building blocks 2, 3, and 4 (re-

dispatch; low- or zero-emitting generation, including

renewables; and demand-side energy efficiency,

respectively.)

---

[50]   CAA section 404(f)(2)(B)(iii)(I).

(IV). EPA Precedent

In the past, the EPA has promulgated rules under CAA section 111(d), in conjunction with CAA section 129, that were based on measures that are similar to some of the measures in the building blocks that EPA proposes as the basis for the regulatory requirements in this rulemaking. For example, the EPA has authorized states to allow large municipal waste combustors to average their emission rates and trade NOx emission credits,[51] and have required the owners of certain waste incineration facilities to take steps to reduce the amount of waste that the facilities combust.[52]

---

[51] *See* "Standards of Performance for New Stationary Sources and Emission Guidelines for Existing Sources; Municipal Waste Combustors," 60 Fed. Reg. 65,387 (Dec. 19, 1995) (trading rules codified in 40 C.F.R. section 60.33b(d)(1)-(2)). EPA also authorized an emission trading program in the Clean Air Mercury Rule. *See* "Standards of Performance for New and Existing Stationary Sources: Electric Utility Steam Generating Units, Final Rule," 70 Fed. Reg. 28,606 (May 18, 2005) *vacated on other grounds by New Jersey v. EPA*, 517 F.3d 574 (D.C. Cir. 2008), *cert denied sub nom. Util. Air Reg. Grp. v. New Jersey*, 555 U.S. 1169 (2009); "Standards of Performance for New Stationary Sources and Emission Guidelines for Existing Sources; Municipal Waste Combustors," 60 Fed. Reg. 65,387 at 28616-24, (Dec. 19, 1995).

[52] *See, e.g.*, Standards of Performance for New Stationary Sources and Emission Guidelines for Existing Sources: Hospital/Medical/Infectious Waste Incinerators, 62 Fed. Reg. 48,348, 48359 (Sept. 15, 1997); Standards of Performance for New Stationary Sources and Emission Guidelines for Existing Sources: Commercial and Industrial

(V). Other considerations

It should also be noted that a number of commentators in the private sector and academia have indicated support for interpreting the term, "system of emission reduction" to incorporate measures such as re-dispatch, renewable energy, and demand-side energy efficiency.[53] Some stakeholders have as well.[54]

Solid Waste Incineration Units, 65 Fed. Reg. 75338, 75341 (Dec. 1, 2000).

[53] *See* Nordhaus R., Gutherz I., "Regulation of CO_2 Emissions from Existing Power Plants Under section 111(d) of the Clean Air Act: Program Design and Statutory Authority," *Environmental Law Reporter,* 44: 10366, 10384 (May 2014) ("strong arguments for" interpreting "system" to include measures such as the addition of new zero-carbon generating capacity and increases in end-user energy efficiency); Sussman R., "Power Plant Regulation Under the Clean Air Act: A Breakthrough Moment for U.S. Climate Policy?" *Virginia Environment Law Journal*, 32:97, 119 (2014) ("EPA would seem to have discretion to define 'system' to include any mix of strategies effective in reducing emissions."); Konschnik K., Peskoe A., "Efficiency Rules: The Case for End-Use Energy Efficiency Programs in the Section 111(d) Rule for Existing Power Plants," *Harvard Law School Environmental Law Program – Policy Initiative* 4 (March 3, 2014) (EPA is authorized to "consider[] … the entire [electricity grid] system when setting performance standards."); Monast J., Profeta T., Pearson B., Doyle J., "Regulating Greenhouse Gas Emissions From Existing Sources: Section 111(d) and State Equivalency," Environmental Law Reporter, 42: 10206, 10209 (March 2012) ("Demand-side energy-efficiency programs and renewable energy generation may fit within the section 111 framework, however, because both reduce the utilization of power plants …. According to this reasoning, emission reductions are occurring within the source category, because of changes in generation at the power plant.").

[54] Ceronsky M., Carbonell T., "Section 111(d) of the Clean Air Act: The Legal Foundation for Strong, Flexible & Cost-

In addition to the just-discussed reasons why interpreting the term, "system of emission reduction" to include those measures is a reasonable interpretation under the CAA, that interpretation also is reasonable as a matter of policy, as we discuss extensively in the preamble. To reiterate briefly, including those measures is consistent with the industry's long-standing methods of operating to assure reliability at the least cost, how states have more recently reduced non-greenhouse gas air pollutants from the industry, and, how states and segments of the industry have, still more recently, reduced CO_2 emissions.

b. "Best system of emission reduction … adequately demonstrated"

For the reasons described next, the measures in each of building blocks 2, 3, and 4 qualify as components of the

Effective Carbon Pollution Standards for Existing Power Plants," *Environmental Defense Fund,* at 9 (Oct. 2013), *available at* http://www . edf .org/sites/default/files/111-clean_air_act-strong_flexible_cost-effective_carbon_pollution_standards_for_existing_power_plants .pdf ; Doniger D., "Questions and Answers on the EPA's Legal Authority to Set 'System Based' Carbon Pollution Standards for Existing Power Plants under Clean Air Act Section 111(d)," *NRDC [Natural Resources Defense Counsel] Issue Brief* (Oct. 2013); "Comments of the Attorneys General of New York, California, Massachusetts, Connecticut, Delaware, Maine, Maryland, New Mexico, Oregon, Rhode Island, Vermont, Washington, and the District of Columbia on the Design of a Program to Reduce Carbon Pollution from Existing Power Plants" (Dec. 16, 2013).

"best system of emission reduction ... adequately demonstrated." As noted elsewhere, the D.C. Circuit has interpreted the BSER as "[a]n adequately demonstrated system," and explained that such a system is one that can "be[] shown to be reasonably reliable, reasonably efficient, and ... reasonably ... expected to serve the interests of pollution control without becoming exorbitantly costly in an economic or environmental way."[55] In fact, the measures in the building blocks do meet the criteria established by the Court in the section 111 case law. In addition, the measures are "adequately demonstrated" because they have already been implemented in many states, and because they may be undertaken by the affected EGUs in the regulated markets in which they operate, or may be implemented by the states in the state plans.

i. Criteria for the BSER

The measures in building blocks 2, 3, and 4 meet the criteria for inclusion as components of the BSER because they are individually and together technically feasible, and together they achieve significant emission reductions, are not unreasonably costly, and will promote the development and implementation of technology improvements

---

[55] *Essex Chemical Corp. v. Ruckelshaus*, 486 F.2d at 427.

for continued emission reductions.[56] The bases for these conclusions are discussed in detail in the preamble and briefly summarized below.

Building block 2, which entails substituting generation at higher emitting units (fossil fuel-fired steam generating units) by shifting to generation at lower-emitting affected sources (existing NGCC units) is technically feasible because the NGCC units are already providing electricity to the grid and have sufficient capacity to generate the additional amount of electricity that would substitute for the generation at fossil fuel-fired steam generating units. Re-dispatch is already widely used (usually more in response to fuel price signals than as a CO_2 mitigation measure), including by companies that own both coal- and natural-gas-fired EGUs. It should be noted that there are several mechanisms through which states could cause re-dispatch to occur. First, a state could use its permitting authority to impose limits on the hours of operation (or emissions) of individual steam generating units over a given time period. Second, a state

---

[56] As noted above, we are proposing to determine BSER as the combination of all four building blocks, and because we discuss in the preamble the reasons why building block 1 meets the criteria for inclusion in the BSER, and why the BSER is the combination of all four building blocks, we are not further discussing those points here.

could change the relative costs of generation for more carbon-intensive and less carbon-intensive generating units by imposing a cost on carbon emissions. A state could do so through any of several market-based mechanisms. One would be to adopt an allowance-based system. An example is the Regional Greenhouse Gas Initiative, an allowance-based system in which sources purchase allowances in periodic auctions. Another way would be through a tradable emission rate system, under which the state would impose an emission rate on the steam generating unit that the unit could meet only by purchasing the right to average its emission rate with a unit with a lower rate, such as an NGCC unit. Most broadly, an allowance system would provide the greatest incentive for the most carbon-intensive affected sources to reduce emissions as much as possible so as to reduce their need to purchase allowances (or to allow them to sell un-needed allowances), and the same would be true for a tradable emission rate system.

As discussed in the preamble, building block 3, which entails use of new low- and zero-emitting generation, as well as preservation of nuclear capacity that might otherwise be retired, is also technically feasible. The technology for renewable energy is well-established and in use now, and the amount of renewable energy contemplated by the proposal would not impair the reliability of the grid. The nuclear capacity at issue either is already in operation or, in the case of new nuclear capacity under construction, has long been known to grid operators for planning purposes. The measures in building block 3 may be implemented in different ways, including market mechanisms. In particular, markets for renewable energy certificates, which facilitates investment in renewable energy, are already well-established. In addition, as noted above with re-dispatch, an allowance system or tradable emission rate system would provide incentives for sources to reduce their emissions as much as possible, including by substituting their generation with generation from renewable energy.

As for building block 4, as discussed in the preamble, numerous state and utility programs have demonstrated that improvements in demand-side energy efficiency are technically feasible at the levels contemplated in the proposal. An allowance system or tradable emission rate system would provide incentives that promote the measures in building block 4 in the same manner as just discussed for other building blocks.

For the reasons discussed in the preamble, the combination of building blocks 2, 3, and 4, along with building block 1, also meet the criteria to qualify as the

BSER. The level of CO_2 emissions reduction they achieve is significant, which is appropriate because of the severity of the risk to public health and the environment of climate change, and the magnitude of both the amount of emissions reductions needed and the amount of CO_2 emissions from fossil fuel-fired power plants.

In addition, based on the measures in building blocks 2, 3, and 4 combined, the proposed levels of reduced generation are not unreasonably costly for the affected source category or the nation-wide electricity system. These levels do not have adverse effects on the overall energy system. Electricity consumers would continue to have access to the electricity they need under these building blocks, although they would need less energy for the same amount of economic activity as a result of the measures in building block 4. Additionally, the measures in building blocks 2, 3, and 4 would improve the electricity system by reducing its carbon intensity, as well as other pollutants, allowing consumers to get the same amount of electricity for less environmental harm. Together, these measures would also promote the development and implementation of technology that is important for continued emissions reductions.

ii. Basis for "adequately demonstrated" finding

The measures in building blocks 2, 3, and 4 are "adequately demonstrated" because each of the individual measures is adequately demonstrated, and because it has been adequately demonstrated that the measures can be taken in combination with each other in a manner consistent with the criteria for determining the BSER.

The measures in building block 1 are adequately demonstrated because they are based on the real-world experience of individual power plants in recent years, as more fully described in the preamble and a technical support document.

The measures in building blocks 2, 3, and 4 are "adequately demonstrated" because, as discussed in the preamble, due to the integrated nature of the electricity system, they have long been relied on to reduce costs in general, assure reliability, and implement pre-existing pollution control requirements in the least cost manner. As also noted in the preamble, some utilities, states and regions are already relying on these measures for the specific purpose of reducing CO_2 emissions from EGUs.

At the same time, as discussed in the preamble, measures in building blocks 2, 3, and 4 may be undertaken,

71

and in fact have been undertaken, by the affected EGUs themselves, which further indicates that these measures are "adequately demonstrated." To achieve the re-dispatch described in building block 2, operators of the affected fossil fuel-fired steam-generating EGUs may reduce generation, while operators of the affected NGCC units may increase generation to replace that avoided at higher-emitting facilities. Operators of the affected EGUs may invest in, or otherwise acquire power from, the new low- or zero-carbon intensive generation described in building block 3, as well as in many of the demand-side energy efficiency measures described in building block 4.

More specifically, many states maintain a utility regulatory structure under which the utilities that serve end users in the state are vertically integrated, and not only own the EGUs, but often also own renewable energy resources and provide service directly to retail customers. Operators of EGUs, in those circumstances, are well-positioned to undertake the measures in building blocks 3 and 4. In fact, as noted in the preamble, numerous states have already imposed renewable portfolio standards and demand-side energy efficiency requirements on those utilities. As a result, as also noted in the preamble, many companies have already developed integrated resource plans that include re-dispatch from higher-emitting fossil fuel-fired generation to lower-emitting generation, the purchase of renewable capacity or the development of renewable generation assets, and the implementation of demand-side energy efficiency measures.[57]

Other states have de-regulated their electricity markets[58] and as a result, in some instances, the EGUs in those states are merchant generators that sell to the wholesale electricity market. The EPA believes that markets for acquiring renewable energy resources and for delivering demand-side energy efficiency services are sufficiently well-developed that operators of these EGUs could undertake or acquire those measures as well. For example, merchant generators can invest in NGCC capacity, invest in renewable capacity or purchase renewable energy or renewable energy certificates (representations that a certain amount of energy was produced from renewable sources), as well as purchase demand-side energy efficiency services from energy

---

[57] Moreover, in many de-regulated states, forward capacity auctions are used to ensure the ability to meet future demand, and generators may bid into those auctions based on all of their resource portfolio,
including renewable energy assets and demand-side energy efficiency projects. This has encouraged generators to undertake the measures in building blocks 3 and 4.
[58] Some states, such as Ohio, have hybrid model that includes elements of a regulated market and a de-regulated market.

service companies. The fact that the affected sources may themselves implement or invest in the measures in building blocks 2, 3, and 4 -- which, again, reduce their emissions -- supports treating those measures as components of the BSER.

Another reason that the measures in building blocks 2, 3, and 4 should be considered "adequately demonstrated" – and wholly apart from the fact that the EGUs may undertake those measures themselves – is based on the fact that CAA section 111(d)(1)(A) provides, by its terms, that the standards of performance that are based on the BSER must be established by the states in state plans. As a result, emissions reduction measures that the states themselves have the authority under state law to put in place may be considered to be part of the BSER. While EGU owners and operators may effectuate such measures directly or indirectly, the states also have authority to enact measures such as dispatch limitations, renewable portfolio standards that require investment in renewable energy resources, as well as demand-side energy efficiency measures.[59] As noted in the preamble, many states have already done so.[60]

_____

[59] It should be noted that under the portfolio approach to the state plan, discussed in the preamble, the entities

Finally, we note that during the public outreach sessions, stakeholders generally recommended that state plans be authorized to rely on, and that affected sources be authorized to implement, re-dispatch, renewable energy measures and demand-side energy efficiency measures, in order to meet the states' and sources' emissions reduction obligations. The EPA agrees that state plans may include these measures, at least under certain circumstances discussed in the preamble, and that sources may rely on them to achieve required reductions. It is clear that these types of measures are well-accepted by the stakeholders as means to reduce emissions from affected sources. The fact that state plans and sources would be expected to use these types of measures to reduce emissions supports the view that these measures are part of a "system

---

that undertake some of the measures in, for example, building block 4 may not be the affected EGUs. Regardless of which entities undertake the measures in the building blocks, those measures have the effect of reducing CO_2 emissions from fossil fuel-fired EGUs, and therefore each of the building blocks remains part of a "system" of emission reduction for those EGUs.

[60] More than half the states have established renewable portfolio standards (RPS) that require minimum proportions of electricity sales to be supplied with generation from renewable generating resources. More than 20 states have energy efficiency resource standards (EERS) that require utilities to effectuate a certain amount of savings in electricity demand each year or cumulatively. Database of State Incentives for Renewables & Efficiency (DSIRE), http://www.dsireusa.org/summarymaps/index.cfm?ee=0&RE=0.

of emission reduction" for those sources that the EPA may evaluate against the appropriate criteria to determine whether they comprise the "best system of emission reduction … adequately demonstrated."

c. Stakeholder concerns

As noted above, some stakeholders have argued that section 111(a)(1) authorizes the EPA to identify re-dispatch, low- or zero-emitting generation, and demand-side energy efficiency measures (building blocks 2, 3, and 4) as components of the "best system of emission reduction … adequately demonstrated." However, other stakeholders have disagreed that this approach is consistent with CAA section 111(d). According to these latter stakeholders, as a legal matter, the BSER is limited to measures that may be undertaken at the affected electric generating units (EGUs), including on-site controls, activities, or work practices, and cannot include measures that are beyond the affected units. These stakeholders take the position that although efficiency improvements at the affected EGUs may be included in the BSER,the measures in building blocks 2, 3, and 4 are "beyond-the-unit" measures because they are implemented outside of the affected EGUs and outside of the control of their owners or operators.[61] Some stakeholders have also argued that section 111(d)(1) requires that the performance standards established by the states must reflect what is achievable at each existing unit.[62]

As the preamble notes, we welcome comment on these issues. As discussed above, we propose that the provisions of section 111 allow the BSER to include those types of measures. In addition, as discussed above, under our proposed approach, affected sources may themselves

---

[61] "Response of the Utility Air Regulatory Group to EPA's 'Considerations in the Design of a Program to Reduce Carbon Pollution from Existing Power Plants" (Oct. 2013); "Existing Source Performance Standards for Greenhouse Gas Emissions from Electrical Generating Units: Creating a Regulatory Framework Under Clean Air Act section 111(d) – A whitepaper from the Coalition for Innovative Climate Solutions" (Feb. 26, 2014); "Perspective of 18 States on Greenhouse Gas Emission Performance Standards for Existing Sources under §111(d) of the Clean Air Act," included in "Testimony before the Subcommittee on Energy and Power of the House Committee on Energy and Commerce – 'EPA's Proposed GHG Standards for New Power Plants and H.R. __, Whitfield-Manchin Legislation'" (Nov. 14, 2013) (statement of E. Scott Pruitt), http://democrats.energycommerce.house.gov/sites/default/fil es/documents/Testimony-Pruitt-EP-EPA-GHG-Standards-Whitfield-Manchin-Legislation-2013-11-14.pdf. *See* National Climate Coalition, "Discussion Background Paper: Best System of Emission Reduction" (Oct. 16, 2013)("BSER approach that mandates reductions based on actions outside the control of the regulated source would involve legal uncertainty. There is nothing in the CAA that authorizes EPA to issue guidelines that require a standard to be based on something that is outside the fence and outside the control of the source.")
[62] "Response of the Utility Air Regulatory Group to EPA's 'Considerations in the Design of a Program to Reduce Carbon Pollution from Existing Power Plants" (Oct. 2013).

implement the measures included in building blocks 2, 3, and 4, so that those measures are within their control. Moreover, under our proposed alternative approach, the "system of emission reduction" includes reductions in utilization at the affected sources themselves.[63]   It

---

[63]   Commenters have critiqued this "at-the-unit" and beyond-the-unit" distinction as follows:

> There is an argument that the at-the-unit/beyond-the-unit distinction is not a meaningful one. Specifically, it could be argued that the distinction between at-the-unit and beyond-the-unit measures is largely artificial, because all of the emission reductions under consideration—whether from at-the-unit measures (e.g., fuel-switching or efficiency upgrades) or from beyond-the-unit measures—are, in fact, emission reductions at or from electric generating units on the interconnected electric grid. For example, neither the addition of renewable generation nor the reduction of end-user demand directly reduces atmospheric emission of CO_2; rather these measures permit fossil EGUs to reduce their own output and emissions. It can be argued that all of the systems of emission reduction here contemplated—whether they involve end-use energy efficiency, displacing high-emission generation with lower emission generation, fuel-switching, heat-rate improvements, etc.—are effectively at-the-unit measures that ultimately reduce emissions solely from regulated EGUs. If energy-efficiency programs, added renewable energy, and redispatch from higher emitting facilities to lower emitting facilities are viewed as at-the-unit systems of emission reduction, the at-the-unit/beyond-the-unit distinction arguably becomes irrelevant—at least from a legal perspective.

Nordhaus R., Gutherz I., "Regulation of CO_2 Emissions from Existing Power Plants Under section 111(d) of the Clean Air

78

should also be noted that, as discussed above, the re-dispatch measures in building block 2 are limited to affected sources. In addition, we discuss below that the performance standards that the states may establish under our approach meet the requirements of section 111(d)(1) and section 111(a)(1) because they would reflect the degree of the required emission performance level (which, in turn, is based on the BSER, as the EPA has applied it to the state's sources) that the state assigns to the affected EGUs. Thus, the proposed approach and alternative described next respond to these stakeholder concerns.

4. Second approach: Heat rate improvement measures inbuilding block 1 plus reduced utilization at levels commensurate with building blocks 2, 3 and 4

The EPA is also proposing an alternative approach to the BSER: heat rate improvements (building block 1) combined with reduced utilization in specified amounts of the affected fossil fuel-fired EGUs, commensurate with the amount of low- and zero-emitting generation and avoided generation in building blocks 2, 3, and 4. The reasons why the measures in building block 1 qualify as a component of this approach to BSER, and the reasons why the combination

---

Act: Program Design and Statutory Authority," *Environmental Law Reporter,* 44: 10366, 10383 n. 133 (May 2014).

of building block 1 with the reduced generation qualify as the BSER are the same as discussed above in connection with the first approach to BSER and in the preamble, and will not be discussed further in this subsection 4. Instead, this subsection will discuss the reduced generation component of this second approach to BSER.

Under this approach, the measures in building blocks 2, 3, and 4 would not be components of the system of emission reduction but instead would serve as bases for quantifying the reduced generation (and therefore emissions) at affected EGUs, and assuring that the amount of reduced generation meets the criteria for the "best" system that is "adequately demonstrated" because, among other things, the reduced generation can be achieved while the demand for electricity services can continue to be met in a reliable and affordable manner. Specifically, the amount of generation from the increased utilization of NGCC units would determine a portion of the amount of the generation reduction component of the BSER for affected fossil fuel-fired steam EGUs; and the amount of generation from the use of expanded low- and zero-carbon generating capacity that could be provided, along with the amount of generation from fossil fuel-fired EGUs that could be avoided through the promotion of demand-side energy efficiency, would determine a portion of the amount of the generation reduction component of the BSER for all affected EGUs.

For the reasons discussed below, reduced generation in the specified amounts is a "system of emission reduction," and meets the criteria to qualify as the "best" that is "adequately demonstrated."

a. "System of emission reduction"

Reduced generation is encompassed by the terms of the phrase "system of emission reduction" in CAA section 111(a)(1), as a matter of *Chevron* step 1, because, in accordance with the above-discussed definition of "system," reduced generation is a "set of things" – which include reduced use of generating equipment and therefore reduced fuel input – that the affected source may take to reduce its CO_2 emissions.

If the phrase "system of emission reduction" is not considered clear by its terms, then it may reasonably be interpreted under *Chevron* step 2 to include reduced generation, for several reasons. First, Congress has recognized reduced utilization in several contexts as a method to reduce air pollution. Beginning with the 1970 CAA Amendments, Congress has recognized that SIPs under CAA section 110, in order to assure reductions in NAAQS pollutants to meet attainment requirements, may need to impose emission limits on industrial sources that those sources could meet only by retiring.[64] Similarly, in adopting CAA section 112, which directed the EPA to promulgate emission standards for sources of hazardous air pollutants to a level of stringency that provides an "ample margin of safety to protect the public health,"[65] Congress was clear that the standards could be sufficiently stringent so that "effectively, … a plant would be required to close because of the absence of control techniques."[66]

---

[64] *See* CAA section 110(g) (authorizing temporary emergency suspensions of SIP revisions if needed to prevent the closing of a source of air pollution), enacted as CAA section 110(f) in the 1970 CAA Amendments; 116 Cong. Rec. 42384 (Dec. 18, 1970), *reprinted in* 1970 Legislative History, vol. 1, at 132–33 (statement of Sen. Muskie) (discussing criteria for sources to receive compliance date extensions). Similarly, Congress recognized that to achieve the NAAQS, it was necessary to reduce emissions from motor vehicles, and that an important method of doing so could be restricting the use of motor vehicles in urban areas that were already highly polluted. For this reason, Congress included in the 1970 CAA Amendments authorization for SIPs under section 110 to include "transportation controls." CAA section 110(a)(2)(B), as approved in the 1970 CAA Amendments. Sen. Edmund S. Muskie (D–ME), who led the proponents for the Amendments in the Senate, explained that for some areas to attain the NAAQS, "[c]entral city use of motor vehicles may have to be restricted." 116 Cong. Rec. 42384 (Dec. 18, 1970), *reprinted in* 1970 Legislative History, vol. 1, at 132 (statement of Sen. Muskie).
[65] CAA section 112(b)(1)(B), as enacted in the 1970 CAA Amendments.
[66] 116 Cong. Rec. 42385 (Dec. 18, 1970), *reprinted in* 1970 Legislative History, vol. 1, at 133 (statement of Sen. Muskie). Sen. Muskie added that the emission standards set

Congress's recognition that closing plants is a method of reducing pollution necessarily encompasses reduced utilization as a system of reducing pollution. As a result, it is reasonable to interpret the term "system of emission reduction," which Congress mandated as the basis for controls on section 111(d) air pollutants, to include reduced production.

Other examples of reduced utilization as a means of reducing emissions to comply with CAA requirements are found in settlement agreements between the EPA and fossil fuel-fired EGUs to resolve alleged violations of the CAA new source review (NSR) requirements. These agreements typically allow the EGUs to choose one of several means to comply with their emission reduction obligations, including retiring units.[67]

Reduction of, or limitation on, the amount of generation is already a well-established means of reducing emissions of pollutants in the electric sector,

---

by the EPA "could include emission standards which allowed for no measureable emissions," *id.*, which further suggests that, as a practical matter, the standards could result in reduced production.

[67] *See, e.g.,* Consent Decree, *USA v. Wisconsin Power and Light Co.,* Civil Action No. 13-cv-266 (WWi.DC), at 18, section IV, available at http://www2.epa.gov/sites/production/files/documents/wisconsinpower-cd.pdf

notwithstanding the fact that as a practical matter, some facilities may have to operate, or remain available, to ensure system reliability. For example, reduced generation by higher-emitting sources is one of the compliance options available to, and used by, EGUs to comply with the Clean Air Act acid rain program in CAA title IV, as well as the transport rules that we refer to as the NOx SIP Call[68] and the Clean Air Interstate Rule (CAIR).[69] Reduction in generation is also a possible means by which an EGU can achieve compliance with its requirements under RGGI.

b. "Best system of emission reduction … adequately demonstrated"

Reduced generation in specified amounts meets the criteria to be the "best" system of emission reduction that is "adequately demonstrated." Reduced generation is technically feasible due to the source's ability to limit its own operations. Moreover, because the amount of reduced generation may be substituted with the building block 2, 3, and 4 measures for increased generation from low- or zero-emitting sources and increased demand-side energy efficiency, that amount may be determined with precision and may be accomplished in a manner that assures the reliability of the electricity grid.

---

[68] 63 FR 57356 (Oct. 27, 1998).
[69] 70 FR 25162 (May 12, 2005).

Specifically, through this reduced generation approach, the amount of emission reduction achieved is appropriate, as discussed above. In addition, the cost of the levels of reduced generation are reasonable for the affected source category and the nation-wide electricity system and do not jeopardize reliability. This is because the measures in building blocks 2, 3, and 4 are already in widespread use in the industry, and it is reasonable to expect that these measures will develop to achieve the levels proposed as part of this approach and thereby ensure an adequate and reliable supply of electricity. Moreover, reduced generation from fossil fuel-fired EGUs and its replacement through the measures in building blocks 2, 3, and 4 is consistent with trends in the energy sector and offer promise to reduce the carbon intensity of the system over the near- and long-term. This approach also promotes the development and implementation of technologies that are important for continued emissions reductions by increasing the demand for those technologies. This is because of the interconnected nature of the electrical grid and the fungibility of electricity, which allows decreases in utilization at one facility to be seamlessly offset by increased utilization elsewhere (building blocks 2 and 3) or by decreased demand (building block 4), and thereby makes reduced utilization a viable approach for emissions reductions by EGUs. Further, this fungibility increases over longer timeframes with the opportunity to invest in infrastructure improvements, and as noted elsewhere, this proposal provides an extended state plan and source compliance horizon. Thus, this approach is consistent with the case law, which authorizes the EPA to determine the BSER by "balanc[ing] long-term national and regional impacts," and by "using a long-term lens with a broad focus on future costs, environmental and energy effects of different technological systems…."[70]

Reduced generation in those amounts is also "adequately demonstrated." As noted above and discussed further in the preamble, the measures in building blocks 2, 3, and 4 are already in widespread use in the industry. At the levels proposed, they have the technical capability to substitute for reduced generation at some or all affected EGUs at reasonable cost. The NGCC capacity necessary to accomplish the levels of generation reduction proposed for building block 2 is already in operation or under construction. Moreover, it is reasonable to expect that the incremental resources reflected in building blocks 3 and 4

---

[70] *Sierra Club v. Costle*, 657 F.2d 298, 331 (D.C. Cir. 1981).

will develop at the levels requisite to ensure an adequate and reliable supply of electricity at the same time that affected EGUs may choose or be required to reduce their CO_2 emissions by means of reducing their utilization. There are several reasons for this. First, the affected sources themselves could invest in new renewable energy resources and demand-side energy efficiency, as discussed in the preamble.[71] Second, the states, as part of their plans, have mechanisms available to put these substitutes in place: they could establish requirements or incentives that would result in new renewable energy and demand-side energy efficiency programs, as also discussed in the preamble.[72] Third, as also discussed in the preamble, regional entities in the electricity system can accommodate these substitutes.

Most broadly, with respect to the measures in building blocks 2, 3, and 4, provided there is sufficient lead time

---

[71]It should be noted that in light of the low current and projected near term prices for natural gas, market forces may lead investors to choose to build new NGCC units, rather than new renewable resources. This result would not call into question the technical feasibility of a BSER that included reductions in fossil fuel-fired generation by the amount of a specified amount of new renewable resources. This is because under these circumstances, the fossil fuel-fired generators could still reduce their generation without causing reliability or other problems in the electric power system.

[72] The nuclear generating capacity reflected in building block 3 is already in operation or under construction.

for planning, mechanisms are in place in both regulated and deregulated electricity markets to assure that substitute generation will become available and/or steps to reduce demand will be taken to compensate for reduced generation by affected EGUs. These mechanisms are based on, among other things, the integrated nature of the electricity system coupled with the availability of capacity in existing NGCC units, the growing institutional capacity of entities that develop renewable energy and demand-side energy efficiency resources, and the ability of system operators and state regulators to incentivize further development of those resources.

7. *Re-dispatch and sources in the regulated source categories.*

As described in the preamble, building block 2 consists of reductions in generation from fossil fuel-fired steam generating units, and corresponding increases in generation by NGCC units. The amount of this re-dispatch is the amount that the steam generating units may reduce, and that NGCC units may increase, up to an average of 70% capacity utilization of the NGCC units.

Accordingly, this component of the BSER involves two sets of affected sources. The first (the steam generating units) decreases their emissions. The second (the lower-

88

emitting NGCC units) may increase their emissions if increased operations are necessary to ensure the ongoing reliability of the integrated electricity system, of which both sets of source are a part, as emissions and generation reduction is occurring at steam generating units and net reductions are being achieved. Both these sets of sources are affected sources because they are in source categories that are covered by this rulemaking. As noted in the preamble, the fossil fuel-fired steam generating boilers are in a source category that the EPA listed under CAA section 111(b) in 1971, and the NGCC units are in a source category that EPA listed in 1979. The NGCC units (as well as the steam-generating units) are subject to reduction requirements through other components of the BSER, specifically, building blocks 3 and 4 (low- and zero-emitting energy and demand-side energy efficiency, respectively). In addition, as noted in the preamble, the EPA is co-proposing to combine the two source categories into a single source category, covering fossil fuel-fired EGUs.

8. Building blocks 2, 3, and 4: intra-state and inter-state compliance

In this section, we discuss the issue of whether CAA section 111(d) limits the EPA to applying the re-dispatch

89

component (building block 2) of the BSER, based on the assumption that each state will comply with that component on a purely intra-state basis, or instead, whether the EPA could base building block 2 on an assumption that the states will comply with that component through the interstate region with which they share the grid.

As the preamble describes, in evaluating building block 2, we have assumed that each state would implement it on a state-by-state basis, without relying on a multi-state regional grid. In particular, we have assumed that each state would increase generation of its own NGCC units to as close to the proposed average 70% capacity utilization as possible, given the amount of generation from in-state fossil fuel-fired steam generating units, and we assumed the corresponding amount of reduction in generation from those steam generators. We have determined the costs of that re-dispatch, and propose to find that they are reasonable.[73] Because we know that dispatch systems operate over multi-state regions, however, we have also determined the costs of the re-dispatch if each state that is part of a multi-state grid implements re-dispatch by taking into account the multi-state grid in which it operates.

---

[73] It should be noted that we also evaluated region-wide re-dispatch, for which the costs are less.

We found that based on the intra-state approach, some states could not increase their average NGCC unit utilization to 70% because they have limited fossil fuel-fired steam generation. In contrast, based on the region-wide approach, more of the states could increase their average NGCC utilization to 70%. In addition, the costs of the intra-state approach are demonstrably higher than the costs of the region-wide approach. In fact, we expect that because all of the lower-48 states, with the exception of Texas, are part of a multi-state, regional grid each state's implementation of building block 2 would, as a practical matter, necessarily occur on an interstate, and not an intrastate, basis.

CAA section 111(d)(1), by its terms, applies requirements on a state-by-state basis. It requires that "each State shall submit to the Administrator a plan" that includes standards of performance as well as implementing and enforcing measures. Further, it allows "the State in applying a standard of performance to any particular source under a [state] plan" to take into consideration factors such as the source's remaining useful life.

These provisions raise the issue of whether section 111(d) may be interpreted so that the re-dispatch component of the BSER may be applied on the assumption that each state would implement that component on a purely intra-state basis, or whether section 111(d) may be interpreted so that the re-dispatch component may be applied on the assumption that each state would implement through the operation of the interstate grid in which it participates. This issue may also apply to building blocks 3 and 4.

C. Application of the BSER; achievability of the emissions standards

1. Introduction and Overview

In this subsection C, we discuss our interpretation of the CAA sections 111(d)(1) and 111(a)(1) requirements that each state must develop a plan that establishes for "any existing source"[74] "standards of performance," which are defined as emission standards that "reflect the degree of limitation achievable through the application of the best system of emission reduction ... adequately demonstrated." We explain why our state-wide approach to applying the BSER and the emission standards that result from the state plan process we require are consistent with these section 111 provisions.

---

[74] It should be recalled that although in this subsection C. we refer to "any existing source" or "each existing source" in the state, or we use similar terms, CAA section 111(d) applies to only those existing sources that would be covered by a section 111(b) standard if they were newly constructed or if they modified or reconstructed.

These provisions make clear that an important aspect of the state's establishment of the standards of performance is "the application of" the BSER. In this rulemaking, the EPA is proposing to apply the BSER for affected EGUs on a statewide basis. The statewide approach also underlies the required emission performance level, which is based on the application of the BSER to a state's affected EGUs, and which the suite of measures in the state plan, including the emission standards for the affected EGUs, must achieve overall. The state has flexibility in assigning the emission performance obligations to its affected EGUs, in the form of standards of performance -- and, for the portfolio approach, in imposing requirements on other entities -- as long as, again, the required emission performance level is met.

This state-wide approach both harnesses the efficiencies of emission reduction opportunities in the interconnected electricity system and is fully consistent with the principles of federalism that underlie the Clean Air Act generally and CAA section 111(d) particularly. That is, section 111(d) achieves the emission performance requirements through the vehicle of a state plan, and provides each state significant flexibility to take local circumstances and state policy goals into account in determining how to reduce emissions from its affected sources, as long as the plan meets minimum federal requirements.

For convenience, we set out the requirements of CAA sections 111(d)(1) and 111(a)(1) here: under CAA section 111(d)(1), the state must adopt a plan that "establishes standards of performance for any existing source." Under CAA section 111(a)(1), a "standard of performance" is a "standard for emissions … which reflects the degree of emission limitation achievable through the application of the best system of emission reduction … adequately demonstrated." The EPA proposes to interpret these provisions as set forth in this sub-section.

The first step is for the EPA to determine the "best system of emission reduction … adequately demonstrated." As discussed at length elsewhere, the EPA is proposing two alternative BSER. The first is the measures in building blocks 1 through 4 combined. This includes operational improvements and equipment upgrades that the coal-fired steam-generating EGUs in the state may undertake to improve their heat rate by, on average, six percent and increases in, or retention of, zero- or low-emitting generation, as well as measures to reduce demand for generation, all of which, taken together, displace, or avoid the need for, generation from the affected EGUs. This BSER is a set of measures that impacts affected EGUs as a group. The alternative approach to BSER is building block 1 combined with reduced utilization from the affected EGUs in the state as a group, in the amounts that can be replaced by an increase in, or retention of, zero- or low-emitting generation, as well as reduced demand for generation.

After determining the BSER, the EPA then applies the BSER to each state's affected EGUs, on a state-wide basis. Building block 1 is applied to the coal-fired steam-generating EGUs on a statewide basis; building block 2 is applied to increase the generation of the NGCC units in the state up to certain amounts, and decrease the amount of generation from steam-generating units accordingly; and the measures in building blocks 3 and 4 are applied to reduce, or avoid, generation from affected EGUs on a state-wide basis. Under the alternative formulation of the BSER, the total amount of reduced generation from the affected EGUs in the state, associated with the measures in building blocks 2, 3, and 4, is determined on the basis of each state's affected EGUs as a group.

This statewide approach to applying the BSER is consistent with the CAA section 111(a)(1) definition of "standard of performance," which, as quoted above, refers to "the application of the [BSER]," for the purpose of determining "the degree of emission limitation achievable," but does not otherwise constrain how the BSER is to be applied. As a result, we, as the administering agency, have discretion under Chevron step 2 to fashion an interpretation that is a reasonable construction of the CAA provisions.[75] Similarly, the implementing regulations give the EPA broad discretion to identify the group of sources to which the BSER is applied. The regulations provide that the EPA "will specify different emission guidelines or compliance times or both for different sizes, types, and classes of designated facilities when costs of control, physical limitations, geographical location, or similar factors make subcategorization appropriate."

In this rulemaking, the EPA is applying the BSER to the affected EGUs in each state as a group. As we have noted, for this industry, a state-wide approach harnesses the efficiencies of emission reduction opportunities in the interconnected electricity system, including the

---

[75] *Chevron U.S.A. Inc. v. NRDC*, 467 U.S. 837, 842–844 (1984).

opportunities to reduce emissions from all affected EGUs through reasonable cost, lower-emitting replacement generation. Accordingly, under the implementing regulations just quoted, it is "appropriate" to apply the BSER to the affected EGUs in each state as a group.

As part of applying the BSER, the EPA, to return to provisions of CAA section 111(a)(1), calculates the "emission limitation achievable through the application of the [BSER]." In this rulemaking, we refer to this amount as the state goal. As noted, the EPA expresses the state goal in the emission guidelines as an emission rate.

The state must develop a state plan that achieves the state goal, either in the form of an emission rate, as specified for the state in the emission guidelines, or a translated mass-based version of the rate-based goal. We refer to the state goal, in the form used by the state as the foundation of its plan, as the required emission performance level.

As part of its state plan, the state must establish "standards of performance" for its affected EGUs. To do so, the state may consider the measures the EPA identified as part of the BSER or other measures that reduce emissions from the affected EGUs. Moreover, the state has the flexibility to establish emission standards in the degree of stringency that the state considers appropriate.[76] The primary limitation on the state's flexibility is that the emissions standards applied to all of the state's affected EGUs -- and, in the case of states that adopt the portfolio approach, the requirements imposed on other affected entities -- taken as a whole, must be demonstrated to achieve the required emission performance level. In addition, the state may make the emission standards for any of its affected EGUs sufficiently stringent, so that the standards and any requirements imposed on other affected entities (if relevant), taken as a whole, achieve a level of emission performance that is better than the required emission performance level. See CAA section 116, 40 CFR 60.24(g).[77]

Under these circumstances – that the emission standards that the state establishes for its affected EGUs

---

[76] Looked at another way, through our proposal, consistent with the EPA's authority in determining the BSER to subcategorize sources on the basis of costs and other factors, see 40 CFR 60.22(b)(5), the state has the opportunity in effect to subcategorize its sources on the basis of their costs and other considerations associated with their position in the interconnected electricity grid, and to assign responsibilities for achieving the emission performance level accordingly.

[77] By comparison to state implementation plans (SIPs) under CAA section 110, although section 111(d) state plans differ from SIPs in that the latter are designed to achieve a NAAQS, section 111(d) plans that are designed to achieve a required emission performance level incorporate many of the same flexibilities as SIPs.

and any other requirements for the other affected entities, as relevant, taken together, are at least as stringent as necessary to achieve the required emission performance level for the state's affected EGUs – each emissions standard that the state adopts for each of its affected EGUs will meet the definition of a "standard of performance" under CAA section 111(a)(1). Specifically, the "standard of performance" for each source will constitute, to return to the provisions of CAA section 111(a)(1), "a standard for emissions which reflects [that is, embodies, or represents][78] the degree [that is, the portion] of emission limitation achievable through the application of the [BSER]" [that is, as noted above, the required emission performance level for all affected sources in a state]. That "degree" or portion of the required emission performance level is, in effect, the portion of the state's obligation to limit its affected sources' emissions that the state has assigned to each particular affected source. An emissions standard meets this definition of the term "standard of performance" regardless of whether it is part of a plan that adopts the portfolio approach (in which

---

[78] See Oxford Dictionary of English (3rd ed. 2010 (online version 2013)) (defining "reflect" as, among other things, "embody or represent (something) in a faithful or appropriate way").

case, the standard will reflect a relatively smaller part of the emission performance level) or one that imposes the plan's emission limitation obligations entirely on the affected EGUs (in which case, the standard will reflect a relatively larger part of the emission performance level).[79]

These proposed interpretations of the provisions of CAA sections 111(d)(1) and (a)(1) are fully consistent with the EPA's overall approach in this rulemaking to determining and applying the BSER and identifying the appropriate level of emission performance for the affected EGUs. As noted, this approach entails applying the BSER on a state-wide basis and, based on the BSER, identifying the emission performance level that each state must achieve, so that each state may then assign responsibilities for achieving that performance level among its sources. As

---

[79] The EPA's approach may also be characterized as (i) determining the BSER for the affected EGUs, (ii) establishing as the emission guideline the standard for emissions that the affected EGUs in the state can achieve on average through the application of the BSER, and (iii) as part of the emission guideline, authorizing each state to establish as the applicable standard for each affected EGU, the standard that the state considers appropriate and that when totaled with the standards established for the other EGUs (and as may be adjusted to account for the portfolio approach, if that approach is adopted by the state) is at least as stringent as the average standard in the emission guideline. As noted in the accompanying text, a state has many ways to establish standards that meet the CAA requirements, including, for example, following the BSER or authorizing emission rate averaging or trading.

noted, this approach is fully consistent with the interconnected nature of the electricity system and with the principles of federalism that form part of the foundation of the Clean Air Act, and that find expression in section 111(d) through its provisions implementing the required emission controls through the vehicle of state plans. We also note that, as part of our proposal for BSER, applying the "best system of emission reduction … adequately demonstrated" on a statewide basis in this manner is consistent with interpreting the term "best" to include those principles of federalism. That is, one reason why each of our proposed two alternative approaches for BSER qualifies as the "best" system is that, in effect, each can be implemented in an efficient manner by a state – through its obligation to assure achievement of the emission performance level that is based on the BSER -- which may mean assigning greater responsibility for emission limitations to some affected EGUs than to others.

It should be emphasized that each state has many options for assigning the emission limitation obligations among its affected sources.[80] For example, the state could

---

[80] One of the advantages of the flexibility states have under the EPA's approach is that state officials may utilize their knowledge of the electricity sector in their

impose emission standards that directly flow from the BSER.

Under these circumstances, the state may assign to

different affected sources emission standards with

different levels of stringency because the state will have

determined that those standards are consistent with the

extent to which the low- or zero-emitting generation in

building blocks 2, 3, and 4 will displace the source's

generation and thereby lower the source's emissions. The

state may establish a relatively less stringent emission

standard for a source that the state considers will not

have much of its generation displaced than the state may

for a source that the state considers will have more of its

generation displaced.[81] The state could base this approach

on the recognition that the increased zero- and low-

emitting generation displaces generation from affected

sources in different amounts, depending on the affected

state and of the entities involved in fashioning the
standards of performance and other requirements.
[81] It should be noted that if the state wished to pattern
the emissions standards after the way that the source was
affected by the BSER, the state would also need to consider
the extent to which the source can implement the heat rate
improvements in building block 1, but for purposes of
simplifying this example, we will set that consideration
aside. It should also be noted that this example assumes
that the state, in assigning emission rates to its sources,
credits reductions in emissions due to reductions in
generation against the emission rate.

sources' costs and on other factors, such as transmission line capacity.

In addition, the state could authorize emission trading as part of the emission standards for affected sources. Under these circumstances, if an affected source's emissions level was higher than the standard the state established for it, the source could achieve the standard by purchasing additional emission rights through the trading program.

It bears emphasis that each state has flexibility in establishing the standards of performance for its existing sources as long as, on a state-wide basis, those standards (and, in the case of the portfolio approach, any other permissible measures in the state plan) achieve the state's required emission performance level. This flexibility is in keeping with the nature of the BSER that we have determined and the state-wide manner in which we have applied it to each state's existing sources. This flexibility is also consistent with the interconnected nature of the electricity system, through which the fossil fuel-fired EGUs are connected to, and affect, each other, and are all affected by other sources of generation.

Finally, it should be noted that states retain authority under CAA section 116 and 40 CFR 60.24(g) to impose standards of performance that, cumulatively, are

more stringent than the emission performance level.